AF843278

Essential Oil Extraction, Scalability Concerns, and Process Intensification

This book provides insights into essential oil extraction research with a special interest in the scalability concerns of each extraction method and their techno-economic concerns. It includes advancements and prognosis for the optimal extraction of essential oils from laboratory to pilot, then to industrial scale. It overviews and identifies sustainable options and proposes improvements to the currently available technologies in essential oil extraction.

Features:

- Focuses on extraction technology of essential oils and scalability concern of essential oil extraction.
- Presents robust elucidation on essential oil production, current market value, and demand.
- Reviews feasibility studies and consideration of essential oil production processes.
- Recent advancements in oil extraction methodologies and process intensification techniques for optimizing nutraceutical recovery.
- Emphasizes process optimization control parameters, and addresses environmental sustainability concern in essential oil extraction.

This book is aimed at graduate students, researchers, and professionals in separation and purification processes, process intensification, and agricultural feedstocks.

Essential Oil Extraction, Scalability Concerns, and Process Intensification

A Comprehensive Guide

Edited by
Abiola Ezekiel Taiwo,
Olalere Olusegun Abayomi and
Bernardo Castro Dominguez

CRC Press
Taylor & Francis Group
Boca Raton London New York

CRC Press is an imprint of the
Taylor & Francis Group, an **informa** business

Designed cover image: Shutterstock

First edition published 2026
by CRC Press
2385 NW Executive Center Drive, Suite 320, Boca Raton FL 33431

and by CRC Press
4 Park Square, Milton Park, Abingdon, Oxon, OX14 4RN

CRC Press is an imprint of Taylor & Francis Group, LLC

© 2026 selection and editorial matter, Abiola Ezekiel Taiwo, Olalere Olusegun Abayomi and Bernardo Castro Dominguez

Reasonable efforts have been made to publish reliable data and information, but the author and publisher cannot assume responsibility for the validity of all materials or the consequences of their use. The authors and publishers have attempted to trace the copyright holders of all material reproduced in this publication and apologize to copyright holders if permission to publish in this form has not been obtained. If any copyright material has not been acknowledged please write and let us know so we may rectify in any future reprint.

Except as permitted under U.S. Copyright Law, no part of this book may be reprinted, reproduced, transmitted, or utilized in any form by any electronic, mechanical, or other means, now known or hereafter invented, including photocopying, microfilming, and recording, or in any information storage or retrieval system, without written permission from the publishers.

For permission to photocopy or use material electronically from this work, access www.copyright.com or contact the Copyright Clearance Center, Inc. (CCC), 222 Rosewood Drive, Danvers, MA 01923, 978-750-8400. For works that are not available on CCC please contact mpkbookspermissions@tandf.co.uk

Trademark notice: Product or corporate names may be trademarks or registered trademarks and are used only for identification and explanation without intent to infringe.

ISBN: 978-1-032-55555-3 (hbk)
ISBN: 978-1-032-63416-6 (pbk)
ISBN: 978-1-032-63418-0 (ebk)

DOI: 10.1201/9781032634180

Typeset in Times
by Apex CoVantage, LLC

Contents

About the Editors

Abiola Ezekiel Taiwo is a chemical engineer, lecturer, and researcher, a recipient of the National Research Foundation (NRF) and The World Academy of Science (TWAS) African Renaissance Doctoral Fellowship grant at Cape Peninsula University of Technology, South Africa between 2016 and 2019. He was awarded the Green Chemistry best price in the 2017 Elsevier Foundation – TWAS Sustainability Case Studies Competition award in Rio de Janeiro Brazil. He is currently a research fellow at the Faculty of Engineering, Mangosuthu University of Technology, Durban, South Africa. His research areas of specialization and interests are artificial intelligence, bioprocessing, data science, food science, nutritional studies, modeling and simulation, process optimization, and renewable energy. He has received invitations to participate in national and international conferences and public lectures. He has published in reputational journals and book chapters and serves as a reviewer of many high-impact journals and book review processes.

Olalere Olusegun Abayomi is an experienced researcher with demonstrated multidisciplinary expertise in chemical engineering science, food biotechnology, bioprocess engineering, and analytical chemistry. He finished his PhD from the University Malaysia Pahang with a specialization in chemical and natural resources engineering. Before that he obtained his master's degree in industrial and production engineering from the University of Ibadan, Nigeria. He had his first degree in chemical engineering from the Ladoke Akintola University of Technology, Ogbomoso, Nigeria. He has authored over 60 research articles in Scopus and Web of Science and has been cited more than 1,000 times, with an h-index of 20 and i10-index of 34. He currently works as a lecturer at the University of Greenwich, Central Avenue, Chatham Maritime, Kent ME4 4TB, United Kingdom. His research interest among many others includes bioprocess engineering, biochemical engineering separation/extraction process, optimization, new product development, and biopharmaceutical engineering. He is an active reviewer for many journals, such as the *Journal of Cleaner Production, Journal of Food Process Engineering, Journal of Food Measurement and Characterization, The Journal of Analysis, Food Science and Biotechnology*, and *Journal of Food Process Engineering*. He is presently an associate editor at the *Jundishapur Journal of Natural Pharmaceutical Products.*

Bernardo Castro-Dominguez is an associate professor (tenured) in chemical engineering and lead of the Smart Process Engineering Lab at the University of Bath (UK). His research has focused on the integration of digital technologies for materials discovery and process design and optimization. He has more than 30 publications in scientific journals (e.g., Nat Comm), three book chapters and one patent, and manages a research portfolio of more than £1.3 million. He is the core member of the Institute for Sustainability, the EPSRC Continuous Manufacturing Hub (CMAC), and the deputy director of the Centre for Digital, Design and Manufacturing at the University of Bath. He has worked as a researcher at the University of Limerick (Ireland) and Worcester Polytechnic Institute (USA). He was recipient of the prestigious Monbukagakusho Scholarship to pursue his PhD studies at the University of Tokyo (Japan) in the area of chemical systems engineering. He holds a BSc and MSc in chemical engineering from the University of Utah (USA).

Contributors

Jacky S. Bouanga Boudiomboa
Dairy Processing Technology Centre (DPTC)
Bernal Institute, University of Limerick
Limerick, V94 T9PX, Ireland

Anamika Chauhan
Department of Home Science
ChamanLal Mahavidhyalya, Landhora
Haridwar 247664, Uttarakhand – India

Sapna Jarial
Department of Agricultural Economics and Extension
Lovely Professional University
Punjab-144411, India

Chiazor Faustina Jisieike
School of Engineering Technology
Department of Mechanical Engineering
Federal Polytechnic Oko, Nigeria

Harjinder Kaur
Department of Horticulture
Lovely Professional University
Punjab-144411

Rupesh Kaushik
Department of Horticulture
Lovely Professional University
Punjab-144411

Sanjeev Kumar
Department of Horticulture
Lovely Professional University
Punjab-144411

Paul Musonge
Faculty of Engineering, Mangosuthu University of Technology
Durban, South Africa

Vizelle Naidoo
Department of Chemical Engineering
University of Johannesburg
Johannesburg, South Africa

Maduka Augustine Nwufo
School of Engineering Technology
Department of Mechanical Engineering
Federal Polytechnic Oko, Nigeria

Ifeyinwa Obiora-Dimson
Department of Electronic and Computer Engineering
Nnamdi Azikiwe University
Awka, Nigeria

Anthony Okoji
Department of Chemical Engineering
University of Delta
Agbor, Delta State, Nigeria

Olalere Olusegun Abayomi
School of Engineering, University of Greenwich
Chatham Maritime, Kent, United Kingdom

Olawale P. Olatidoye
Department of Food Technology
Yaba College of Technology
P.M.B 2011, Yaba, Nigeria

Samuel Olugbenga Olunusi
Department of Chemical Engineering
UMPSA Universiti Malaysia Pahang
Al-Sultan, Malaysia

Nor Hanuni Ramli
Department of Chemical Engineering
UMPSA Universiti Malaysia Pahang
Al-Sultan, Malaysia

Rishen Roopchund
Department of Chemical Engineering
University of South Africa/
University of Johannesburg
South Africa

Ajay Saroha
Department of Horticulture
Lovely Professional University
Punjab-144411

Naadhira Seedat
Department of Chemical Engineering
University of Johannesburg/
University of Pretoria
South Africa

Abiola Ezekiel Taiwo
Faculty of Engineering
Mangosuthu University of Technology
Durban, South Africa

Banele Vatsha
Research Centre for Synthesis and Catalysis
Department of Chemical Sciences
University of Johannesburg
Auckland Park 2006
Johannesburg, South Africa

Ab Waheed Wani
Department of Horticulture
Lovely Professional University
Punjab-144411

Preface

There are over 3,000 essential oils recognized today, with around 300 of them being commercially important, primarily in the pharmaceutical, culinary, domestic, and cosmetic industries. Due to the few nutraceutical benefits and market demands, there is a growing interest in its study. Though many researchers have documented different methods of extraction of essential oil from different sources, most of their reports do not go into enough detail to cover all the different fields of specialization that are involved in the process, taking into consideration the various scientific and technological various methods of extraction into a single book. It is important to note that optimal extraction of essential oil at an industrial scale is an interdisciplinary issue and as such requires inputs from many experts such as agriculturists, engineers, scientists, and economists. Thus, the book's primary contribution is twofold: it provides precise background information about the nutraceuticals' (i.e., nutritional and medicinal) properties and emerging technologies in harnessing their nutraceutical benefits from different sources.

A major shortcoming of many books in this field is their failure to incorporate the diverse specializations involved in essential oil production. This book takes a multidisciplinary approach, bringing together scientific, technological, and economic perspectives from researchers in science, engineering, and management. It provides valuable insights into the potential impact of various extraction techniques on the quality and market value of essential oils. What sets this book apart is its interdisciplinary perspective, making it a valuable resource for a wide range of readers.

Dr Abiola Ezekiel Taiwo
Faculty of Engineering, Mangosuthu University of Technology, Durban, South Africa.

Dr Olalere Olusegun Abayomi
School of Engineering. University of Greenwich, Medway, United Kingdom

Associate professor Bernardo Castro-Dominguez
Department of Chemical Engineering, University of Bath, Claverton Down, United Kingdom

Acknowledgments

We extend our heartfelt gratitude to all the authors who contributed to this book. Your dedication, expertise, and scholarly efforts have been instrumental in shaping the content and quality of this work. Without your valuable contributions, this book would not have been possible.

We also deeply appreciate the invaluable role of our reviewers, both for the book proposal and for each individual chapter. Your meticulous review, constructive feedback, and commitment to excellence have significantly enhanced the scholarly rigor and coherence of this work.

A special note of appreciation goes to our editorial team for their unwavering support and dedication. We acknowledge the diligent efforts of our esteemed editors **Dr. Abiola Ezekiel Taiwo**, **Dr. Olalere Olusegun Abayomi**, and **Associate professor Bernardo Castro-Dominguez**, whose keen editorial insights and careful attention to detail have refined and polished the manuscript to its final form.

Furthermore, we express our sincere gratitude to our publishing editors, particularly **Dr. Gagandeep Singh** and **Jahnavi Vaid**, for their steadfast support in bringing this book to publication. Your guidance and expertise in the publishing process have been invaluable.

Finally, we would like to acknowledge the time, effort, and commitment of everyone who played a role, directly or indirectly, in the successful completion of this book. Your collective contributions have made this work a reality, and we are truly grateful.

Thank you.

1 Essential Oil's Current Market Value and Demand

Abiola Ezekiel Taiwo, Olalere Olusegun Abayomi, Paul Musonge, Anthony Okoji and Chiazor Faustina Jisieike

1.1 INTRODUCTION

The use of essential oils for medicinal purposes dates back at least a thousand years. The term "essential oil" is used to describe concentrated oils derived from plants that contain the "essence" of the plant (Petruzzi, 2024; Scott, 2024). It is possible to extract essential oil from nearly any plant matter, from flowers to tree bark, through separation techniques such as distillation, supercritical fluid extraction, microwave, or cold pressing, depending on the technical and economic implication of the producer (Olalere et al., 2022). Figure 1.1 shows that essential oils are no longer primarily used therapeutically but rather for aromatherapy, perfume, food processing, incense, and as biological potentials (Sharma et al., 2023). Essential oils have been shown to provide a number of health benefits, such as lowering stress, improving mood, reducing pain, supporting health and well-being, and being used as nutraceuticals (Petruzzi, 2024; Olalere et al., 2024). However, it is important to maintain the greatest quality and consumer welfare while producing essential oils in compliance with the best industrial practices (Olaniran et al., 2023). Essential oils are strong, concentrated compounds that, if used improperly, can be poisonous. Tea tree oil, lavender oil, and rosemary oil are some of the most often utilized essential oils.

1.2 ESSENTIAL OIL'S MARKET SIZE AND PREDICTIONS

According to Petruzzi (2024), Europe dominates the essential oil market, followed by Asia-Pacific and North America. Figure 1.2 shows that the global essential oils market is expected to grow at a compound annual growth rate (CAGR) of 7.6% from 2024 to 2030 (US$23.74 billion). As a result, industries such as food and beverage, personal care, and cosmetics have increased their demand. The market for flavors is expected to grow in the near future as more beer and other alcoholic beverages are consumed. Grand View Research (2024) reports that essential oils

DOI: 10.1201/9781032634180-1

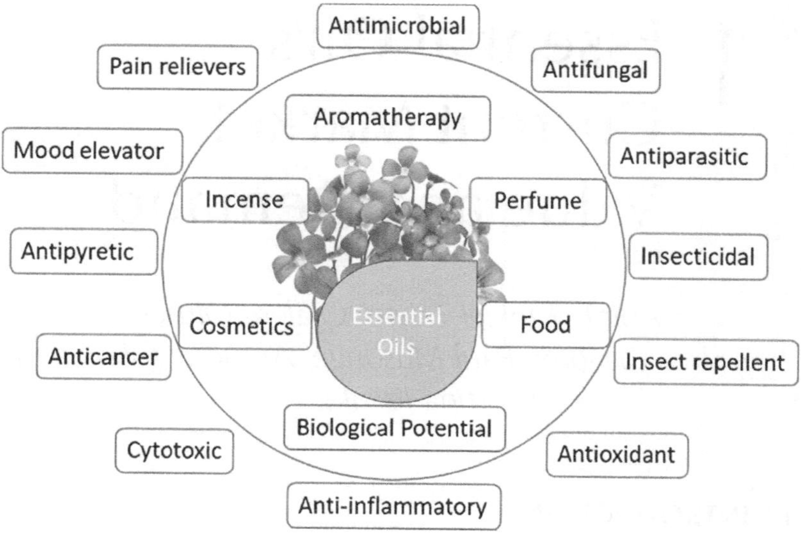

FIGURE 1.1 Essential oil multipurpose use and advantages.

Source: Copyright obtained from Elsevier (Sharma et al., 2023).

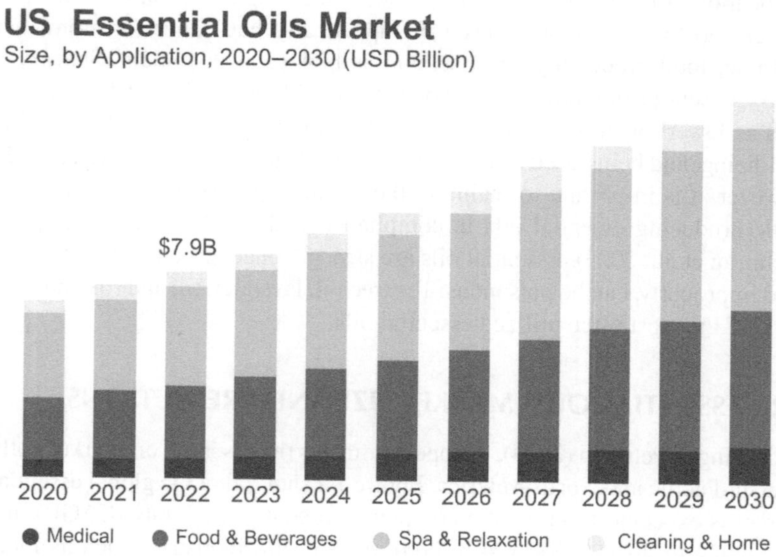

FIGURE 1.2 Essential oils market prediction.

Source: www.grandviewresearch.com.

contain a broad range of antifungal, antibacterial, anti-inflammatory, and antiviral properties. It is worth noting that these oils don't have any side effects similar to most conventional medications and drugs. It is expected that such factors will drive the market's growth. Aside from traditional medicines and drugs, many conventional products are also developed using pleasant-smelling aroma compounds, mainly for chemicals that emit unwanted odors. The health benefits and lower side effects of essential oils are also causing a shift away from most chemicals, drugs, and medicines (Ramsey et al., 2020; Grand View Research, 2024).

1.3 ESSENTIAL OIL MARKET FOCUS AND INTERNATIONAL OIL PLAYERS

The market is expanding quickly and is in a high growth stage. The market for essential oils is distinguished by a high level of innovation because of advancements in technology in the manufacturing process at different points along the value chain. Additionally, there is a tendency to use biotechnology and green chemistry in the development of new products (Grand View Research, 2024).

The major companies in the essential oils market are also actively engaging in mergers and acquisitions to fortify their positions and increase their regional footprints. Leading market participants have used this strategy to strengthen their market positions due to the rapid rise of important end-use industries. Additionally, the market is coming under more and more regulatory scrutiny. For the production, harvesting, and processing of flowers and aromatic crops, international organizations like the European Union (EU) and the United Nations (UN) have established particular rules and regulations (Grand View Research, 2024). There are nomany alternatives to the essential oil business. Essential oils are now being used in place of traditional chemicals in cosmetics due to the growing demand for chemical-free products. Competition, however, can emerge as a result of novel substances or new technologies. Traditional distillation methods are probably going to be replaced by chemical synthesis and biotechnology. The market for essential oils is heavily influenced by end-user concentration (Grand View Research, 2024). The market is anticipated to expand as a result of the rising demand for edible and drink flavors and fragrances as well as natural personal care products made with essential oils and other scent compounds.

1.3.1 METHOD OF EXTRACTION INSIGHTS

The market with the largest revenue share in 2023 was dominated by the distillation method. In aromatic compounds, the increased use of distilled essential oils is responsible for the expansion. When it comes to the manufacture of essential oils, distillation has been chosen since it is the most affordable and cost-effective extraction method (Grand View Research, 2024). Since the process can be conducted near the location of the production plant, steam distillation is popular since it requires little capital. According to Adeyi et al. (2023), there are numerous alternative extraction techniques, such as carbon dioxide extraction, cold press extraction, heat-assisted extraction, and many more. Some disadvantages of these extraction techniques include low yield, toxic solvent residues, long extraction times, and loss of volatile compounds (Stratakos and Koidis, 2016). According to the Grand View Research Report from 2024, the global essential oil market is currently worth about $9 billion, and it is expected to grow significantly over the next several years, with a CAGR of 8–10%. The main drivers of this expansion are the growing consumer preference for natural and organic products as well as the increased attention being given to health and wellness.

1.4 CURRENT ESSENTIAL OIL MARKET DYNAMICS AND CHALLENGES

Citrus oils, mint oils, floral oils, and spice oils are among the market segments for essential oils, each of which serves a variety of purposes in personal care, food and beverage, and aromatherapy. North America leads the market in terms of market share due to high consumer awareness and well-established distribution networks (Scott, 2024). The United States continues to be a major consumer of essential oils because of the growing demand for natural and organic products, as well as the use of essential oils in personal care and wellness products. Canada also contributes to the market growth, as consumers there are becoming more interested in natural ingredients in cosmetics and medications.

The demand for natural cosmetics and health supplements is also growing in Europe, especially in nations like France and Germany. Essential oils are also quite popular in Europe, where demand is highest in nations like France, Germany, and the United Kingdom. France is well known for its extensive history of producing essential oils, especially lavender and rose oils, and aromatherapy. The demand for premium, certified organic essential oils is driven by the strict standards governing product quality and safety in the European market (Scott, 2024).

In Asia-Pacific, people are becoming more interested in natural cures and well-being, particularly in developing markets like China and India. The market for essential oils is expanding significantly in the Asia-Pacific area due to the growing demand for natural and herbal goods in nations like Australia, Japan, China, and India. India is a significant producer of essential oils like citronella, sandalwood, and jasmine, because of its wide variety of plants. Since aromatherapy is becoming more and more popular and essential oils are used in traditional Chinese medicine, China is becoming a major player in the market (Scott, 2024).

1.5 CONSUMER DEMAND, QUALITY MEASURE, AND REGULATORY BODIES IN ESSENTIAL OIL MARKET

1.5.1 Consumer Health and Wellness

The use of essential oils for therapeutic purposes, or aromatherapy, has grown increasingly. In line with the global well-being behavior, there is a significant market trend toward the use of essential oils for tranquility, stress reduction, mood enhancement, and other therapeutic uses. Around the world, the market for healthier goods and services is rising along with interest in leading better lives. Health-promoting food and beverage items have grown in popularity due to a number of factors, including increased food sensitivities, the standardization of health certification processes, and the appeal of socially and environmentally sustainable production (Yu and Watson, 2024). Brands are adopting eco-friendly strategies as a result of consumers' growing concerns about sustainable sourcing. Eco-friendly and sustainable techniques is becoming more and more important to consumers. As a result, there was a rise in demand for essential oils that were sourced ethically and organically. Market share was expected to increase for businesses that could show a commitment to ethical and sustainable business practices (Scott, 2024).

1.5.2 Quality Measure in Essential Oil Market

Despite the market's expansion and demand, there are still issues with essential oil, such as maintaining authenticity and purity since adulteration can erode consumer confidence. According to Pokajewicz et al. (2024), the widespread and immoral practice of adulterating essential oils compromises market integrity and has an impact on both customers and trustworthy providers. The scientific literature describes very few scientific researches, despite the fact that this issue is well known. Pokajewicz et al. (2024) investigated 55 lavender essential oil samples and performed a thorough investigation down to the trace levels, utilizing chiral, quantitative, and chromatographic profiles. Out of the 51 samples that were identified as *Lavandula angustifolia*, 51% were authentic, 6% were cheaper lavandin essential oil, 14% were fortified with artificial additives, and 29% were reported to be tampered with using both synthetic and lavandin essential oil. The authors found that while some costly products were also fake, diluted essential oils were less expensive than authentic ones. Regulations and standards should therefore be in place to measure the quality of essential oils in the market in order to meet customer demand for healthier goods.

1.5.3 Regulatory Hurdles

Regulations that differ from one country to another can make it more difficult to formulate products and access markets. According to Olaniran et al. (2023), there is an urgent need for good agricultural and manufacturing practices that will satisfy the food regulatory standard bodies. Global regulatory agencies like the Food and Drug Administration (FDA), European Food Safety Authority, Food and Agriculture Organization of the United Nations, World Health Organization, and Food Safety and Standards Authority of India handle the addition of foreign antimicrobial and antioxidant agents in the food system differently in each country (Prakash et al., 2024), but they all address consumer safety concerns, particularly not using prohibited additives or excessive amounts of recommended additives (Wu et al., 2021).

Desired information about treaties, laws, and regulations pertaining to botanicals used as flavoring and antimicrobial agents in food and agriculture can be found in the "Food and Agricultural Legislation" and Flavoring Regulation Act (Demyttenaere, 2012; Sharma et al., 2021). Consumers should have access to an online environment for information concerning the use of food additives. For example, the Flavor and Extract Manufacturers Association and the US FDA have maintained a list of essential oils and their biologically active components that are generally regarded as safe, and their use has long been established. According to Prakash et al. (2012), essential oils and their volatile constituents often have favorable LD50 toxicity values, making them the least harmful to human.

1.6 CONCLUSION

The market for essential oils is growing internationally, and consumers are expressing interest in oils from different geographical areas. Oils from specific geographical areas gained prominence due to its distinctive fragrances and cultural significance. Essential oils are used in food flavoring, cosmetics, household cleaning products, and aromatherapy. The usage of essential oils in personal care and aromatherapy has grown significantly. A combination of sustainability concerns, health trends, and the growing demand for natural products is driving the essential oil market's expansion. The market for essential oils is predicted to keep growing as customers look for safe, efficient substitutes to synthetic alternatives, paving the way for sustainable business practices and innovation.

However, the application of nanotechnology to the use of essential oils in food systems has an optimistic outlook to both conventional and technical developments, including increasing the concentration of essential oils in plants, improving volatile management, and creating synergistic combinations.

REFERENCES

Adeyi, O., Adeyi, A. J., Oke, E. O., Okolo, B. I., Olalere, O. A., Taiwo, A. E., Aremu, Oluwole, S., Qwebani-Ogunleye, Tozama, Maphosa, Yvonne, & Ogunsola, Akinola D. (2023). Heat-assisted extraction of phenolic-rich bioactive antioxidants from *Enantia chlorantha* stem bark: Multi-objective optimization, integrated process techno-economics and profitability risk assessment. *SN Applied Sciences*, *5*, 153. https://doi.org/10.1007/s42452-023-05372-1

Demyttenaere, J. C. (2012). The new European Union flavouring regulation and its impact on essential oils: Production of natural flavouring ingredients and maximum levels of restricted substances. *Flavour Fragrance Journal*, *27*(1), 3–12. https://doi.org/10.1002/ffj.2093

Grand View Research. (2024). *Essential oils market size, share and trends analysis report by product (orange, cornmint, eucalyptus), by application (medical, food and beverages, spa and relaxation), by sales channel, by region, and segment forecasts, 2024–2030*. Accessed 6th November, 2024. https://www.grandviewresearch.com/industry-analysis/essential-oils market#:~:text=The%20global%20essential%20oils%20market%20size%20was%20valued

Olaniran, A. F., Taiwo, A. E., Iranloye, Y. M., & Okonkwo, C. E. (2023). The role of good agricultural practices (GAPs) and good manufacturing practices (GMPs) in food safety. In *Food safety and toxicology*. North America, USA: De Gruyter. https://doi.org/10.1515/9783110748345-021

Olalere, O. A., Gan, C. Y., Adeyi, Oladayo, Taiwo, A. E., Adeyi, Abiola, Fawale, Olumide. (2022). Up-scalability and techno-economic perspectives of nonconventional essential oils extraction techniques. *Jundishapur Journal of Natural Pharmaceutical Products,* e122792. https://brieflands.com/articles/jjnpp–122792.html

Olalere, O. A., Gan, C. Y., Taiwo, A. E., Adeyi, O., & Olaiya, F. G. (2024). Essential oils: Sustainable extraction techniques and nutraceuticals perspectives. In Sarkar, T., Pati, S. (eds) *Bioactive extraction and application in food and nutraceutical industries. Methods and protocols in food science*. New York, NY: Humana. https://doi.org/10.1007/978-1-0716-3601-5_15

Petruzzi, D. (2024). *Essential oils market worldwide – statistics and facts*. Accessed 7th November2024.https://www.statista.com/topics/5174/essential-oils/#topicOverview

Pokajewicz, K., Lamaka, D., Wieczorek, P. P., Bortkiewicz, O., & Paluch, E. (2024). Lavender essential oil market integrity: A comprehensive study of commercial Lavandula angustifolia essential oils adulteration and assessment of industrial standards. *Industrial Crops and Products*, *222*, 119744.

Prakash, B., Singh, P. P., Gupta, V., & Raghuvanshi, T. S. (2024). Essential oils as green promising alternatives to chemical preservatives for agri-food products: New insight into molecular mechanism, toxicity assessment, and safety profile. *Food and Chemical Toxicology*, *183*, 114241.

Prakash, B., Singh, P., Mishra, P. K., & Dubey, N. K. (2012). Safety assessment of Zanthoxylum alatum Roxb. essential oil, its antifungal, antiaflatoxin, antioxidant activity and efficacy as antimicrobial in preservation of Piper nigrum L. fruits. *International Journal of Food Microbiology, 153*(1–2), 183–191. https://doi.org/10.1016/j.ijfoodmicro.2011.11.007

Ramsey, J. T., Shropshire, B. C., Nagy, T. R., Chambers, K. D., Li, Y., & Korach, K. S. (2020). Focus: Plant-based medicine and pharmacology: Essential oils and health. *The Yale Journal of Biology and Medicine*, *93*(2), 291.

Scott, E. (2024). *Essential oils market size, share, trends, growth drivers, and regional analysis*. Accessed 6th November 2024. https://www.linkedin.com/pulse/essential-oils-market-size-share-trends-growth-drivers-edward-scott-ftl7f/

Sharma, A., Gumber, K., Gohain, A., Bhatia, T., Sohal, H. S., Mutreja, V., & Bhardwaj, G. (2023). Importance of essential oils and current trends in use of essential oils (aroma therapy, agrofood, and medicinal usage). In *Essential oils* (pp. 53–83). Academic Press.

Sharma, S., Barkauskaite, S., Jaiswal, A. K., & Jaiswal, S. (2021). Essential oils as additives in active food packaging. *Food Chemistry*, *343*, 128403 https://doi.org/10.1016/j.foodchem.2020.128403

Stratakos, A. C., & Koidis, A. (2016). Methods for extracting essential oils. In *Essential oils in food preservation, flavor and safety* (pp. 31–38). Academic Press.

Wu, M., Zhou, Z., Yang, J., Zhang, M., Cai, F., & Lu, P. (2021). ZnO nanoparticles stabilized oregano essential oil pickering emulsion for functional cellulose nanofibrils packaging films with antimicrobial and antioxidant activity. International Journal of Biological Macromolecules, *190*, 433–440.

Yu, M., & Watson, S. (2024). Market overview of health and wellness food products. In Du, X., Yang, J. (eds) *Flavor-associated applications in health and wellness food products*. Cham: Springer. https://doi.org/10.1007/978-3-031-51808-9_1

2 Azeotropic Distillation
Unlocking the Potential of Essential Oil Extraction

Naadhira Seedat, Vizelle Naidoo and Rishen Roopchund

2.1 INTRODUCTION

Distillation is a long thread that weaves through time, with roots originating millennia ago. The rich history of distillation begins approximately 1500 BC in ancient Egypt (Koller et al., 2005; O'Hara, 1998), Greece, India, Persia, and China (Price, 2000; O'Hara, 1998), with the foundation of this significant practice tracing back to the days of Kings, Queens, and Pharaohs. The ancient civilisations of these eras relied predominantly on nature for their daily activities and sought innovative manners to harness an assortment of natural essences for a diverse range of applications (O'Hara, 1998; Price, 2000). What began as alchemists of the time using the principles of heated water to separate plant components to craft balms, perfumes, and remedies were, though unknowingly, the rudimentary practices that paved the way to become later the modern-day chemical engineering technique known as distillation (Price, 2000).

2.1.1 Definition and Significance of Azeotropic Distillation in Essential Oil Extraction

The word "distillation" originates from the Latin word *distillare*, which signifies "to trickle down in minute drops" or "to let fall in drops" (Khalisanni et al., 2014). Distillation is "the process of making a liquid stronger or purer by heating it until it changes to a gas and then cooling it so that it changes back into a liquid", according to the Cambridge dictionary. Scientifically, distillation is defined as the "separation of two components based on their respective relative volatilities" (Richardson et al., 2002). Nonetheless, the fundamental principle of distillation shines brightly as an effective separation technique across several industries (Seedat et al., 2020). The term "azeotrope" refers to a mixture of two or more components in a liquid mixture that has a constant boiling point and exists in a vapour–liquid equilibrium (Rackley, 2017; Salonen et al., 2012). This

 DOI: 10.1201/9781032634180-2

phenomenon makes it difficult to separate these components from one another, leading to the invention of azeotropic distillation. Azeotropic distillation in the context of essential oil extraction refers to distillation procedures that apply heat to plant material, which causes the *in-situ* water to burst the cells releasing the oil (Lucchesi et al., 2007). The oil gets entrained in the *in-situ* water of the plant by azeotropic distillation and further condensed and separated to achieve the oil product (Lucchesi et al., 2007). Azeotropic distillation can be further classified as steam distillation, steam-water distillation, hydro-diffusion distillation, and hydro-distillation (Ibrahim et al., 2021). This chapter will focus on highlighting the operation, advantages, drawbacks, and applications of the three forms of azeotropic distillation processes.

2.2 BRIEF OVERVIEW OF HISTORICAL BACKGROUNDS

The exact commencement of distillation is unknown and surrounded by much debate and speculation. Nevertheless, proof of essential oil utilisation from multiple ancient societies, such as India, Greece, and China, has been extensively recorded (Price, 2000; O'Hara, 1998). The proof suggests that specific extraction techniques were employed to obtain the coveted oils; however, there is limited knowledge about the exact methods used due to the ancientness of the texts in question. Nonetheless, focusing on the primordial tale in ancient Egypt, the use of oils is well documented for embalming bodies of the dead, mummification, perfumery, and tributes to the Gods dating back to as early as 1500 BC (O'Hara, 1998; Koller et al., 2005). According to Van Winkle (1967), one of the first recorded details of batch distillation was during the time of Queen Cleopatra in 50 BC; however, O'Hara (1998) stated that the distillation of essential oils began in Egypt at least in 400 BC. However, there are more ancient texts that pre-date these that detail more historical methods to extract oils and produce remedies, drinks, and perfume, and this insinuates that distillation could have existed and been employed approximately 1,000 to 2,000 years prior (Van Winkle, 1967; Battaglia, 2003; O'Hara, 1998). Primitive distillation practices continued for centuries and were adapted along the way, with a "simple condenser" as a long tube exposed to open air being described in the fourth century (Van Winkle, 1967). It was not until the eleventh century (between AD 980 and 1037) that vast and major improvements were made to distillation processes (Price, 2000), when Persian physician and alchemist Ibn Sina, also known as Avicenna, intervened (Price, 2000; Battaglia, 2003). Sina was regarded as the "Prince of Physicians" and "Aristotle of Arabs" and was described as the first person to perfect steam distillation (Battaglia, 2003; Khalisanni et al., 2014). According to the literature, Sina increased the efficiency of the distillation process by improving the cooling system, resulting in the steam-oil molecules cooling down more rapidly (Price, 2000).

2.3 HISTORICAL EVOLUTION OF ESSENTIAL OIL EXTRACTION

2.3.1 Ancient Methods of Essential Oil Extraction

Essential oils are conventionally extracted from natural plants using water, steam, cold press, dry distillation, and solvent extraction methods (Li et al., 2014; Akdağ & Öztürk, 2019). The chemical composition of these essential oils could vary based on the specific method of extraction used, even if the oil is extracted from the same plant source. The conventional methods of extraction can be influenced by multiple factors, including distillation duration, operating temperature and pressure, and the nature and quality of raw plant materials used in the process (Li et al., 2014). According to Li et al. (2014), conventional distillation techniques may induce thermal impacts or hydrolysis, resulting in the loss of some chemical constituents and the deterioration of some unsaturated components. A similar observation was reported by Akdağ and Öztürk (2019), who noted that although the conventional processes may appear advantageous and more economically feasible, they are capable of producing the same adverse effects of hydrolysis and thermal degradation. These disadvantages have stimulated the intensification and optimisation of existing and novel "green" extraction techniques, which are applied based on the plant materials and the final product quality (Li et al., 2014).

Cedarwood, which was popular for embalming and hygienic purposes, has been described to have been extracted by heating wood in a clay container and covering it heavily with a thick layer of wool. As the process progressed, the wool became saturated with cedarwood essential oil and steam; the wool was then squeezed efficiently, and the resultant oil-water mixture was left to separate in a container (Price, 2000). This method depicts the beginnings of distillation practices, and it can be expected that this method was employed in other flora for oil extraction at the time. Another description of ancient practices includes boiling or steaming plants (flowers) in a closed vessel attached to a thin tube. As the water boils with the contents of the vessel, the oil is released and mixed within the water, where it is decanted into an awaiting container (O'Hara, 1998), as indicated in Figure 2.1.

2.3.2 Transition to More Advanced Distillation Techniques

During the Middle Ages and the Renaissance period, the cultivation of aromatic plants and their natural essences increased to combat the Black Death plague in the fourteenth century, and this brought further attention to distillation techniques (Price, 2000). It was later, at the end of the fourteenth century, that a Swiss physician and alchemist, Paracelsus, stated that distillation released the most sought-after component of plants, which prompted a much higher interest and focus on distillation as an extraction for essential oils (Price, 2000). The sixteenth century ushered in a new era of progression in distillation as the first available books on distillation were published in Germany by physician Jerome Brunswick,

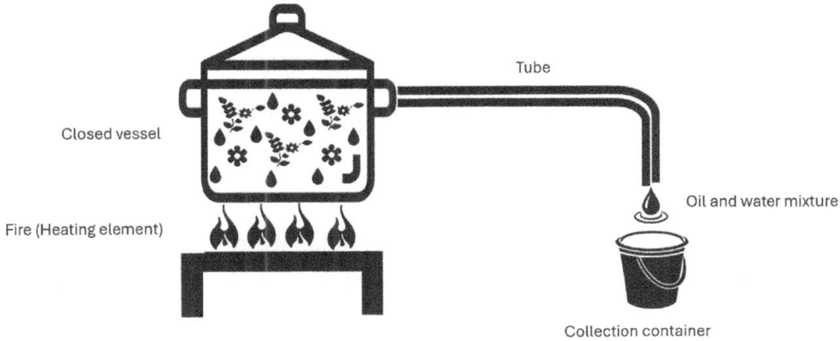

FIGURE 2.1 Graphical depiction of ancient distillation practices.

Source: Image generated by the author.

who is said to have detailed 25 different essential oils, inclusive of nutmeg, myrrh, cinnamon, clove, lavender, and rosemary among others (Battaglia, 2003).

It was in this era that the art of distillation caught the attention of many alchemists of the time (Battaglia, 2003). At this point, distillation was being applied to the separation and recovery of alcohols, water, vinegar, and oils, among other products (Van Winkle, 1967). The progression of distillation continued during this time period, with experimental conversations surrounding the different apparatus configurations gaining pace, resulting in one of the first-ever analytical records of distillation, in which an alchemist separated alcohols according to their respective boiling points (Van Winkle, 1967).

With the rise of chemical engineering in the late eighteenth century, substantial advancements were made in distillation, where steam was used predominantly as a heating agent in distillation processes (Van Winkle, 1967). This further prompted the developments of mathematical models in the nineteenth century, which were applied to the fractional distillation of binary mixtures, bringing focus to more engineering variables such as enthalpy, heat loss, mass and energy balances, rates, reflux, and pressure effects, among others (Van Winkle, 1967). The application of engineering principles and mathematical models was extended to more complex distillation processes such as multi-component distillation, complex column design, computational simulations, and, most prominently, process optimisation, which resulted in the increase in process efficiency known today (Van Winkle, 1967), thus, laying the foundation for the evolution of distillation that extends into the present twenty-first century (Van Winkle, 1967; Schlosser, 2011). The interest in efficient essential oil extraction on a large scale has spurred the development of multiple types of azeotropic distillation and distillation-hybrid models, which will be elaborated in detail in the following sections in this chapter. Thus, what began as primal principles of extraction through the progression of time, the rise of technology, and the shaping through chemical engineering prompted the transformation of small clay pots and cotton wool into fully developed industrial extraction unit operations known and employed using azeotropic distillation processes in the modern-day era, today, to extract essential oil.

2.3.3 Traditional Azeotropic Distillation Techniques

2.3.3.1 Steam Distillation

A. Process Description

Steam distillation has been used to isolate essential oils from plants since ancient times. The process involves passing the steam through the plant materials from the base of the alembic to the top (Li et al., 2014), enabling the plant material to be distilled using steam generated by a boiler (Akdağ & Öztürk, 2019) (see Figure 2.2). Essentially, the plant materials are exposed to steam, prompting the release of their essential oils through evaporation (Stratakos & Koidis, 2016). During the process, the essential oil molecules mix with the rising steam molecules, resulting in a vapour oil–steam mixture. This mixture then flows through a "swan-neck" column and is condensed, ushering in a phase change from vapour to liquid, which is decanted and collected in a Florentine flask (Li et al., 2014; Di Giacomo & Di Giacomo, 2002).

Usually, due to the density of the mixture, the essential oils extracted are lighter or heavier than the water, thus forming a condensed mixture with two immiscible phases that can be separated based on their density difference (Lukose et al., 2023). The temperature and pressure of the steam are the most important parameters affecting the steam distillation process (Kabuba & Huberts, 2009). The

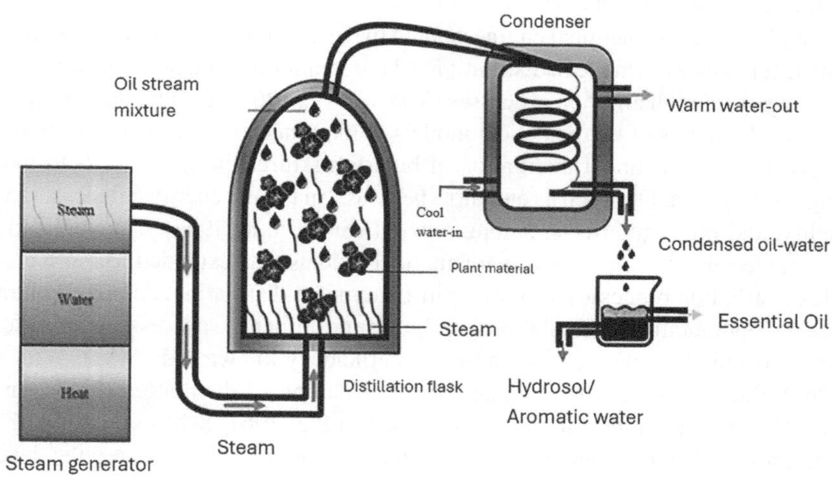

FIGURE 2.2 Diagrammatic representation of the extraction of essential oils using steam distillation.

Source: Image generated by the author.

working principle of this method is structured around the combined vapour pressure of the system, which is expected to be equal to ambient pressure at approximately 100°C. This facilitates the evaporation of the volatile components within the system, with boiling points ranging between 150°C and 300°C to evaporate at approximately 100°C in line with the boiling point of water (Li et al., 2014). Steam distillation can also be performed under different pressures, depending on the difficulties associated with extracting the different essential oils.

According to Stratakos and Koidis (2016), steam and water distillation are essential oil extraction's most common distillation types. About 90% of essential oils are produced using steam distillation (Radwan et al., 2020). These processes can last between one and ten hours (Stratakos & Koidis, 2016). Through these processes, the quantity of essential oils produced depends on the duration of the distillation, type of plant material, temperature, and pressure (Stratakos & Koidis, 2016).

B. Advantages and Limitations

The advantages of steam distillation include the ease associated with controlling how much steam is generated and, thus, the deterioration caused by the heat itself, which can be mitigated or prevented (Öztekin & Soysal, 1998) as a result of the steam generation unit being outside of the distillation unit, and the ambient distillation temperature is maintained below 100°C (Akdağ & Öztürk, 2019). In addition, this method comprises simple instruments and does not require complex technological development (Manzan et al., 2003). This method also allows for the continuous interaction of the plant matter with clean steam to facilitate extraction (Saldaña-Mendoza et al., 2022).

The limitations of steam distillation are associated with the degradation that can potentially occur if the steam flow rate is not carefully regulated (Akdağ & Öztürk, 2019) and the high temperature employed in the process, the formation of artefacts is a possibility and the acidity level of water as the extracted essential oils are exposed to boiling water for extended time frames (Stratakos & Koidis, 2016). These artefacts can lead to the alteration of the compositions in oils, and the potential hydrolysis of esters to alcohols and acids has serious implications (Stratakos & Koidis, 2016). Moreover, some essential oils may require redistillation to eliminate any potential impurities and undesirable constituents that can influence the odour of the oil. This process is known as rectification (Stratakos & Koidis, 2016).

2.3.3.2 Hydro-Diffusion

A. Process Description

According to Kant and Kumar (2022), the hydro-diffusion apparatus comprises a vessel to contain the plant material, a condenser and a hydrosol-essential oil separation unit (see Figure 2.3). The system uses steam to heat the container. The most significant distinction between hydro-diffusion and traditional steam extraction resides in the manner in which steam is integrated into the process

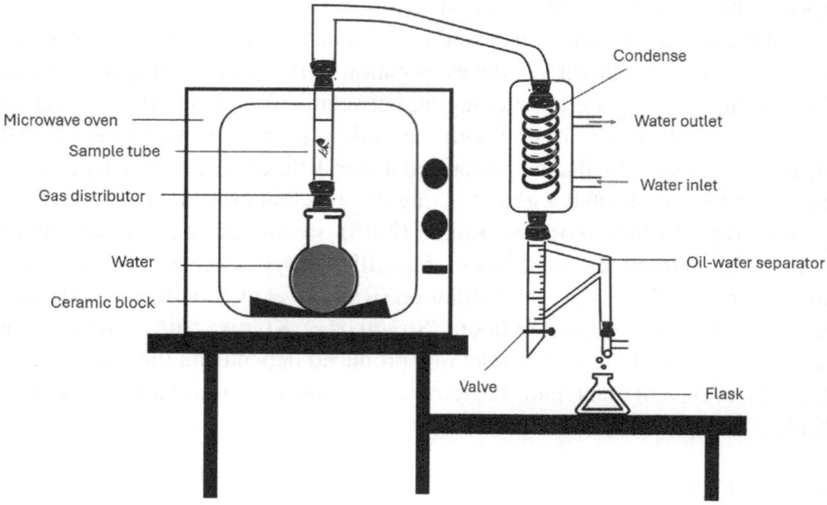

FIGURE 2.3 Illustration representing hydro-diffusion processing of essential oils.

Source: Image generated by the author.

(Battaglia, 2003; Price, 2000). As mentioned earlier in the chapter, steam distillation operates by directing the steam from the bottom of the distillation chamber so that the heated vapours rise through the plant material (Battaglia, 2003; Price, 2000). In the hydro-diffusion process, steam enters from the top of the distillation chamber, allowing it to fall downwards due to gravitational force through the plant material. The hydro-diffusion method is best suited for dried plant samples that are separated below 100°C and can be destroyed at boiling temperatures (Kant & Kumar, 2022).

The study of Yusoff et al. (2013) developed a hydro-diffusion extraction system based on the principles of steam distillation to extract essential oil from kaffir limes to improve the output yield, thus producing essential oil of high quality relative to the conventional steam distillation technique. The hydro-diffusion steam distillation-essential oil extraction apparatus used in the study of Yusoff et al. (2013) consisted of two main segments: the distillation unit and the tray containing the plant material. The system employed an electrical heating element integrated into the distillation unit to generate steam by boiling water. Hydro-diffusion was effectively demonstrated as a novel extraction technique for the extraction of essential oils, in the study of Yusoff et al. (2013). The study described the hydro-diffusion concept as a system in which extract and steam (comprised of essential oil and water) proceed intrinsically downwards to the condenser under the influence of gravity, which differs from conventional steam extraction in which steam moves upwards (Yusoff et al., 2013).

A distillation unit containing ten litres of water was heated using an electrical heater to generate steam, in the study of Yusoff et al. (2013). The kaffir lime peels were located on the plant material tray within the steam hydro-diffusion distillation unit. Thereafter, pressurised steam from above was passed through the kaffir lime peels to vaporise the volatile oils present therein. The material tray was linked to an oil–water separator and condenser. The steam and oil vapour mixture was transferred to the condenser through a small tube, while the volatile components were collected in a container (Yusoff et al., 2013). The separation of the accumulated essential oils from water was undertaken using dichloromethane and sodium anhydrous acid as solvents. The extraction duration was approximately two hours.

B. Advantages and Limitations

According to Yusoff et al. (2013), steam does not undergo complete consumption for extraction in conventional steam distillation processes. As steam moves through the material tray, condensation occurs in the elements, which causes the steam to release its enthalpy of vaporisation. The steam that liberates its vaporisation enthalpy typically returns to the distillation unit, which becomes evident when analysing the water within the distillation unit. However, the reported system of hydro-diffusion steam distillation boasts an enhanced recovery of oil by reducing oil waste contained in boiling water (Yusoff et al., 2013). In conventional steam distillation processes, the residual oil dissolved in the water can give rise to an odour, thus wasting the valuable product contained within the water stream (Masango, 2005). Although re-distilling is often employed to optimise the recovery of essential oil in wastewater, the process is disadvantageous due to high heating and energy costs (Masango, 2005). These concerns are not present in the use of hydro-diffusion, which can also prevent pollution (Abert Vian et al., 2008).

2.3.3.3 Hydro-Distillation

A. Process Description

Hydro-distillation is one of the most traditional (conventional) and commonly used azeotropic distillation methods for retrieving essential oils from organic plant material in addition to steam distillation and water-steam distillation (Nourafcan et al., 2014). According to Rasul (2018), hydro-distillation is a traditional technique because the process involves water and takes place at atmospheric pressure. Water distillation is the most appropriate method to apply to fresh and dry plant materials that are not damaged by heating (Akdağ & Öztürk, 2019). Boiling water is subsequently introduced into the upper portion of the vessel as part of the process in which the plant material is placed, transportation of essential oil by vapour and accumulating in the water due to the density difference when it reaches the Florentine container collection vessel (Akdağ & Öztürk, 2019).

A simple hydro-distillation laboratory apparatus for extracting essential oil is known as the Clevenger apparatus, as shown in Figure 2.4 (Marković et al., 2017). Unlike steam distillation, which uses steam to extract essential oils, hydro-distillation requires directly mixing organic plant material with liquid water in a distillation flask (Li et al., 2014). The resulting solid (plant) and liquid (water) mixture is heated in the distillation flask until the mixture boils at atmospheric

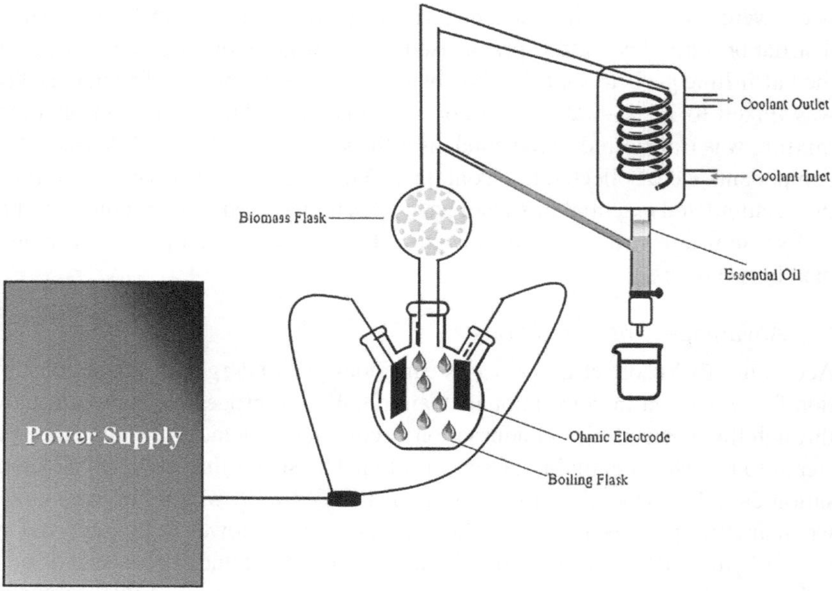

FIGURE 2.4 Hydro-distillation schematic depicting Clevenger apparatus.

Source: Image generated by the author.

pressure. The release of the free bioactive compounds in the plant cells is activated by the heat added to the mixture, whilst the volatile aromatic compounds produce an azeotropic mixture with water (Rasul, 2018). The evaporated oil–water mixture evaporates into a condensation chamber at atmospheric pressure (isotropic operation), in which the mixture condenses. Since oil–water mixtures are immiscible and have a considerable density difference, the oil and water can be separated using a graduated oil-separating tube or a Florentine flask (Marković et al., 2017).

B. Advantages and Limitations

According to Rasul (2018), hydro-distillation has the potential to be a very effective method in comparison to other conventional methods to extract oil from various parts of different plants (Khan & Dwivedi, 2018). The yield of oil from hydro-distillation processes is dependent on the following variables: weight and size of the raw material, the nature of the material and the volume of water mixed with the plant material (Parikh & Desai, 2011):

Three physio-chemical processes, namely hydro-diffusion, hydrolysis, and decomposition by heat, constitute the hydro-distillation process on a molecular level (Rasul, 2018). The constituents of the volatile oil exhibit reduced vulnerability to hydrolysis and polymerisation (Rasul, 2018). The level of dampness at the still's base impacts hydrolysis, while the heat conductivity of the still walls influences polymerisation (Rasul, 2018). One of the advantages of hydro-distillation is that different parts of the various plants can be used to extract oil using the process (Rasul, 2018). In comparison to Soxhlet extraction, maceration, percolation, and decoction, hydro-distillation has higher oil yields (Rasul, 2018). Reduced loss of polar compounds is achieved if the reflux is controlled well (Rasul, 2018). Compared to Soxhlet extraction, the oil quality is easily reproduced (Rasul, 2018). Finally, since no solvents are used, hydro-distillation is economically and environmentally more feasible and sustainable (Rasul, 2018).

The last physio-chemical process of hydro-distillation is the requirement of high extraction temperatures, such that hydro-distillation is only suitable for thermolabile compounds. Hence, the most important factor in the yield of oil is the nature of the raw material (Rasul, 2018). Additionally, the high temperatures and lengthy extraction times may yield chemical alterations in the composition of the oil and the loss of the light key components from the oil (Ferhat et al., 2016). A second disadvantage of the prolonged extraction times is the alteration of the recovered essential oils from the original essence (Li et al., 2014). Other challenges include incomplete extraction, rate of distillation which varies due to difficult control of heat supply, large-scale production required may stills and fuel resulting in uneconomical processes, long extraction times, ester production from hydrolysis of some extracted components due to prolonged action of hot water and contamination of char odours in the essential oil die the contact of organic material in contact with heat source at the bottom of the distillation flask (Rasul, 2018).

The shortcomings listed above and, in particular, alternative energy-saving distillation methods have initiated a path of research to more effective and efficient use of new green processes that reduce the energy required by combining hydro-distillation with microwave, ohmic, and ultrasound-assisted technologies (Ferhat et al., 2016; Li et al., 2014; Gavahian & Farahnaky, 2018).

2.3.3.4 Emerging Hybrid Techniques

A. Introduction to Contemporary Methods

With the progression of engineering, novel green extraction techniques have emerged in recent years and have been used for the extraction of natural essences with a similar or improved quality comparatively to conventional methods, with a reduction in operation units, energy consumption, carbon dioxide emissions, and harmful co-extracts (Li et al., 2014). This section focuses on the emerging hybrid extraction techniques in azeotropic distillation.

B. Mechanisms and Benefits of Hybrid Approaches

1. Microwave-Assisted Steam Distillation According to Li et al. (2014), the microwave is a non-contact source of heat source which can be employed to achieve a greater degree of selective, effective heating. The microwave-assisted steam extraction process is facilitated by electromagnetic non-ionizing radiation, thus changing the thermal state of the irradiated vegetal materials (Saldaña-Mendoza et al., 2022) (see Figure 2.5). The research of Izquierdo et al. (2013) found that microwave radiation adjusts the hydrogen bonds and increases the energy of water molecules, leading to their vibration, which increases the tem-

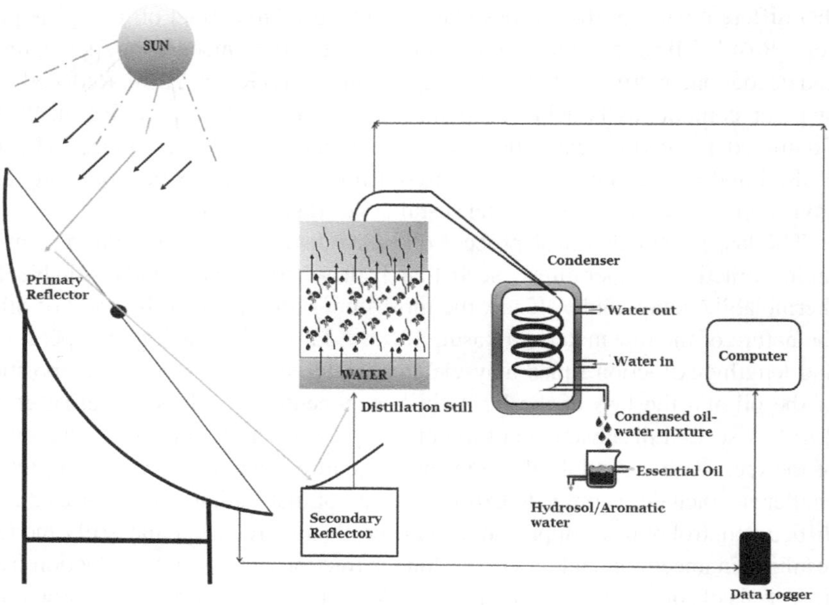

FIGURE 2.5 Graphical depiction of hybrid microwave-assisted steam distillation process.

Source: Image generated by the author.

perature through friction. This sequence, in turn, intensifies the solvent penetration in the matrix due to the migration of ions and dipole rotation (Izquierdo et al., 2013). Chouhan et al. (2019) further indicate that using solvents in the microwave-assisted steam distillation process is optional, as it occurs through the rupture of sebaceous glands in irradiated vegetal matter to release essential oils.

To gain from microwave heating, studies have integrated microwaves with traditional techniques to develop hybrid techniques such as microwave-accelerated steam distillation (Stratakos & Koidis, 2016). Using microwave distillation, essential oil extraction can occur much faster (in minutes instead of hours) with several benefits aligned with sustainable extraction and chemistry principles. This process entails the extraction of vegetal materials within a microwave reactor with or without organic solvents or water. According to Craveiro et al. (1989), the pioneering microwave-assisted essential oil extraction technique was that of compressed air microwave distillation. Using steam distillation principles, compressed air is incessantly introduced into the extractor containing plant materials within water and heated by microwave (Li et al., 2014). In this process, the water and essential oil condensation and separation occur outside the microwave reactor. Notably, the completion of compressed air microwave distillation can occur in just five minutes as opposed to 90 minutes of conventional extraction using steam distillation without any compositional or quality differences in the final product (Li et al., 2014).

There are several advantages of microwave-assisted extraction, such as its exclusive friction-based heating technique, cost efficiency, simplistic design, and satisfactory performance under atmospheric surroundings (Stratakos & Koidis, 2016).

2. *Ohmic-Accelerated Steam Distillation* According to Saldaña-Mendoza et al. (2022), excellent results have been noticed by the addition of ohmic heating-assisted steam extraction (Figure 2.6) in comparison to traditional steam extraction techniques.

Gavahian and Chu (2018) presented a laboratory-scale ohmic-accelerated steam distillation unit for the extraction of lavender essential oil consisting of a 2-L boiling flask with two rectangular-shaped titanium electrodes inserted into the boiling flask. The flask and electrode set-up was connected to a 1-L biomass flask, a modified Clevenger apparatus, a 5-kW power supply, and an industrial chiller. The electrical power of 500 W was supplied from the power supply into 1 L of salted water in the boiling flask through the titanium electrodes to generate steam (Gavahian & Chu, 2018). Saltwater is a more suitable medium than pure water in ohmic-accelerated processes due to its higher electrical conductivity. The

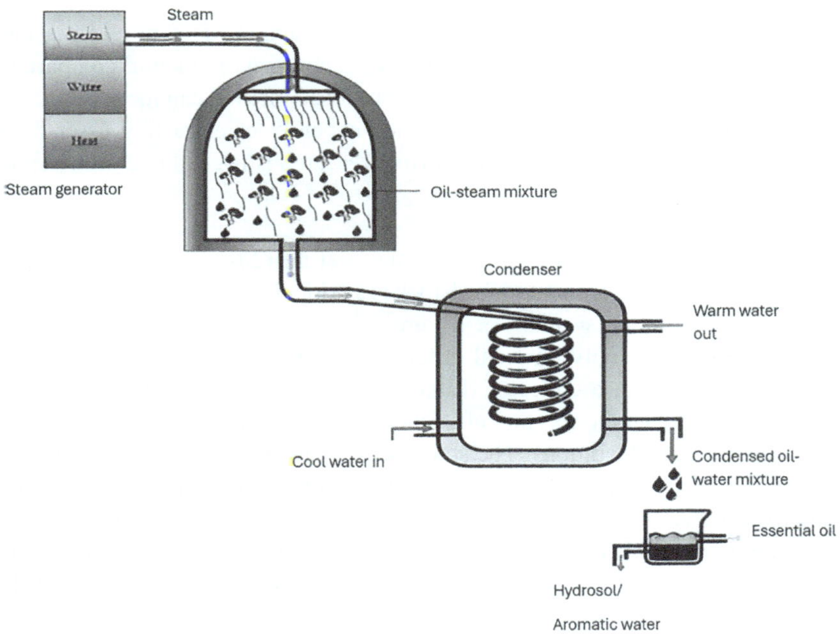

FIGURE 2.6 Graphical depiction of hybrid ohmic-accelerated steam distillation process.

Source: Image generated by the author.

evaporated saltwater from the boiling flask was passed through the 30 g of lavender inflorescences packed into the biomass flask (Gavahian & Chu, 2018). The extracted oil and steam mixture, evaporated from the process, is passed into the traditional Clevenger apparatus section. The evaporated lavender oil and steam were cooled in the Clevenger apparatus by the industrial chiller, which was circulated with a 60% ethanol–water mixture at 4°C. The water–alcohol mixture has a lower freezing point in comparison to water and hence was chosen as the suitable coolant.

According to the study's results (Gavahian & Chu, 2018), the ohmic-accelerated steam distillation reduced extraction time by 55% and energy consumption by 58%. Furthermore, steam distillation advantages, such as continuous contact with the plant material with clean steam, remained in the proposed ohmic-accelerated steam distillation technique. The study also showed that the separation of the ohmic electrodes and plant materials neutralised the adverse effects of potential electrochemical reactions and, hence, electrode corrosion on phytochemicals. Saldaña-Mendoza et al. (2022) further explained that this technology allows the optimised recovery of lavender essential oils as well as preserves the quality of lavender oil compared to the conventional steam distillation method, reducing distillation extraction time and energy consumption to less than 50% (Kant & Kumar, 2021). However, further work is required to explore a greater range of higher electrolyte concentrations, electrical frequencies, and more economical electrode materials.

Ohmic-accelerated steam distillation is advantageous as it provides a more efficient process in terms of energy and extraction time reduction compared to steam distillation for industrial units. The proposed method can increase the production capacity and reduce the production expense for extracting essential oil and distillation industries. However, further research is needed to commercialise the technology.

3. Solar Distillation According to Lukose et al. (2023), selecting a suitable collector based on the amount of feedstock is a critical consideration when designing a solar distillation system (see Figure 2.7). Consideration of factors such as the required productivity, thermal efficiency, required cost, and space must be made when selecting a collector (Lukose et al., 2023). This selection must be done by considering the thermal efficiency, system productivity, required space, and economic considerations. The study of Nannaware et al. (2022) entailed the design of a solar-operated distillation system for the extraction of valuable essentials from vegetal matter with reduced operation costs. The unit incorporates a distillation unit, packed column, mesh grid frame, oil receiver, condenser, energy meter, solar panels, and resistive heating elements (Nannaware et al., 2022). The improved solar-operated distillation system has the capacity to process 20 kg of aromatic crops per batch. Uniform steam generation is achieved within 20 minutes through 4.5-kW-resistive heating elements with solar-powered controllers incorporating the 5-kW photovoltaic panel. Three hours of continuous distillation is possible using the produced steam. The summative power expended by the

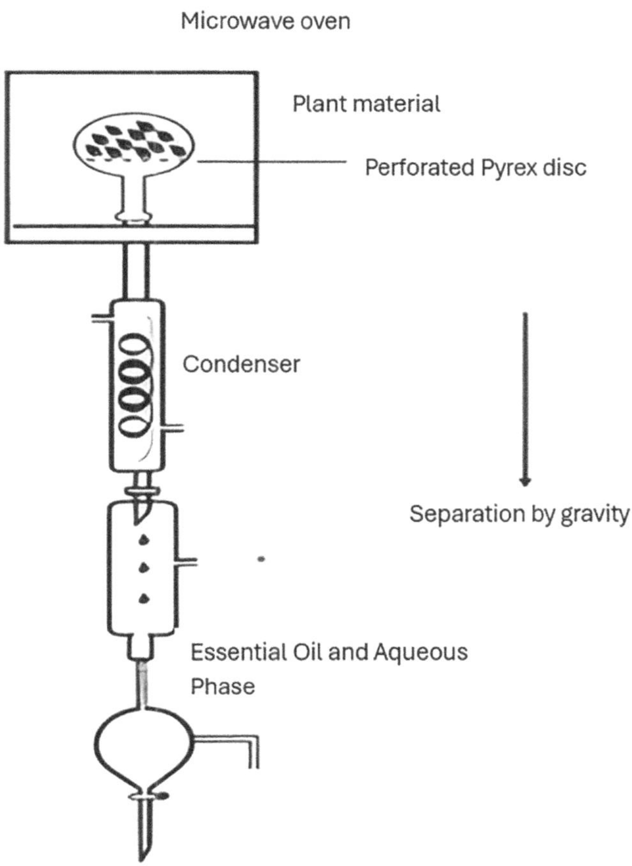

FIGURE 2.7 Graphical depiction of the solar distillation process.

Source: Image generated by the author.

solar-operated distillation unit was calculated to be 13.5 kWh. Analyses indicated that the essential oil attained through solar-operated distillation unit was superior in quality to conventional steam distillation.

Relative to wood-based distillation systems, the quantity of carbon dioxide liberated through the use of the solar-operated distillation system is zero, thus eliminating any adverse environmental effects. The study found that the operating cost was 50% less than the conventional wood-based distillation system. Therefore, the novel technique is economically viable. However, further research is needed to design a concentrator of solar radiation to store solar energy in the event of cloudy weather. The steam distillation unit has several advantages, such as simple operation, consistent steam production, zero carbon dioxide emissions, cost efficiency, high product purity, improved product yield, and minimised likelihood of oil scorching (Nannaware et al., 2022).

Despite these advantages, additional research is required to optimise solar distillation systems. The following aspects were reported by Lukose et al. (2023):

- Integrating solar distillation systems with energy storage systems.
- Improving the yield of essential oil by accounting for specific plant parameters such as the cutting size of herbage and moisture content.
- Modelling critical oil extraction kinetics using solar distillation.
- Determining strategies to improve the thermal efficiency of the system through operation of the system at optimum feedstock amount and using selective coatings at the bottom of the distillation unit.
- Using mathematical models for the prediction of essential oil recovery based on the solar distillation system process parameters.
- Optimising parameters influencing the extraction of essential oil using the solar biomass hybrid system.
- Undertaking chemical analysis of the essential oil product through biomass distillation in solar biomass hybrid systems.

Over the past decade, substantial research has been undertaken to optimise conventional essential extraction methods. New extraction processes must be economical, environmentally friendly, sustainable, efficient (higher yield), with reduced extraction time, and high quality (reduce loss of volatile and other components) (Drinić et al., 2020). Since hydro-distillation is a conventional process, further optimisation is possible using pre-treatment methods such as microwave, ultrasonic, and ohmic to enhance yields and composition of essential oil.

4. Microwave-Assisted Hydro-Distillation Microwave-assisted hydro-distillation has received much recognition since reported in 1980 in the extraction of essential oil over the past few decades (Khan & Dwivedi, 2018). The experimental set-up for a microwave-assisted hydro-distillation unit is a modification of a hydro-distillation process with the addition of a microwave oven attached to the Clevenger apparatus, as shown in Figure 2.8 (Ranitha et al., 2014). The distillation flask of the Clevenger apparatus is filled with the plant material mixed with distilled water and inserted into the microwave oven cavity (Ranitha et al., 2014).

The microwave produces microwave energy, which is a non-ionising form of electromagnetic energy that facilitates the motion of molecules by the movement of ions and rotation of dipoles but does not alter the molecular structure (Ferhat et al., 2016). The required microwave energy range is between a frequency of 300 and 300,000 MHz, a domestic oven provides a frequency of 2 450 MHz (Ferhat et al., 2016). The microwave energy replaces the conventional heating mantle to rapidly deliver energy to the charged still volume of water and plant material mixture to obtain a quick increase in the temperature and boil to induce the process of releasing the oil, thus reducing the time required for extraction (Rezvanpanah et al., 2008). The condenser of the Clevenger apparatus is set outside and above the microwave, where the essential oil extracted is collected (Ranitha et al., 2014).

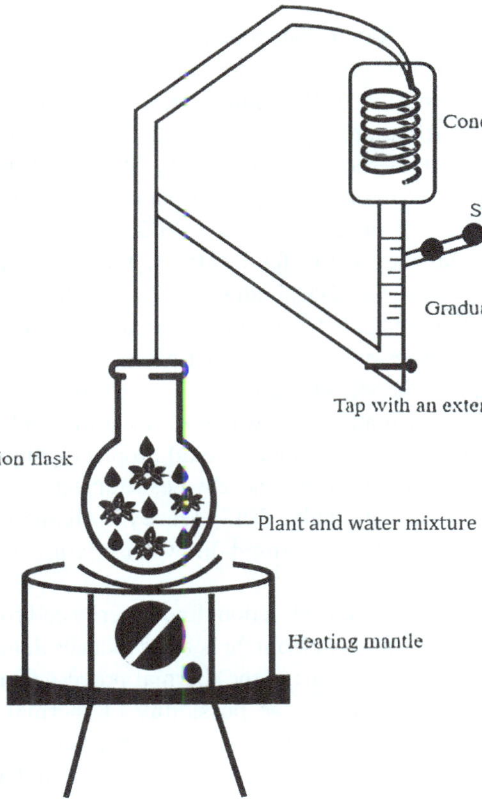

FIGURE 2.8 Diagrammatic representation showcasing microwave-assisted hydro-distillation.

Source: Image generated by the author.

There have been several advantages associated with microwave-assisted hydro-distillation, including short extraction time, higher yields, low cost, increased amount of oxygenated compounds present in the oil, enhanced heating capacity, quicker energy transfer, and environmentally feasible (Ferhat et al., 2006; Abdellatif & Hassani, 2015; Hashemi-Moghaddam et al., 2018; Ranitha et al., 2014). Sample deterioration is one negative effect that can occur through external exposure to microwave radiation (Khan & Dwivedi, 2018). Another hindrance in the limitation of the process is the extent to which the microwave frequency penetrates the material as the power density decreases after the depth of penetration is reached from the power density on the surface of the material (Masum et al., 2019). Thus, after the penetration depth, the effectiveness of heating from the heat source, the microwave, decreases (Masum et al., 2019). Hence, the rate of heat distribution of the material is not uniform across different diameters.

5. Ohmic-Assisted Hydro-Distillation Ohmic-assisted hydro-distillation is a modified process of hydro-distillation that replaces the conventional heating mantle with ohmic heating, which generates thermal energy (Gavahian et al., 2015). A power supply is connected to the ohmic heater. After that, the conventional Clevenger hydro-distillation apparatus is attached to an ohmic heating flask to extract essential oil, as shown in Figure 4.9. Ohmic heating is when the alternating electrical current passes through the plant material to generate internal heat within the plant cells (Gavahian et al., 2012). The electric current generates heat through internal energy transfer (electric to thermal energy by the Joule effect) within the plant material (Gavahian et al., 2012). The rapid heating is achieved within seconds to minutes but relies on electrical conductivity, which dictates the heating rate and applies primarily to water-based systems due to ionic mobility (Gavahian et al., 2012). Hence, saltwater is used as the resistant medium. Oil droplets are low-conductivity materials in water–oil mixtures and cannot be directly heated through ohmic processing. Thus, the oil droplets within the plant material are exposed to rapid heating through nearby heat transfer from the continuous water phase (Gavahian & Farahnaky, 2018). Once the oil and water mixture vapourises into the condenser, the condensed oil–water mixture is separated using a Florentine flask or oil separator.

The main advantage is the accelerated extraction time experienced from ohmic heating. Additionally, it has been shown that ohmic heating reduces damage to the oil and material from heat and fouling during the thermal process (Gavahian & Farahnaky, 2018). The main drawback is the possibility of thermal runaway, which needs to be factored into the process design of the condenser section of the ohmic-assisted hydro-distillation process. Electrode corrosion should also be an

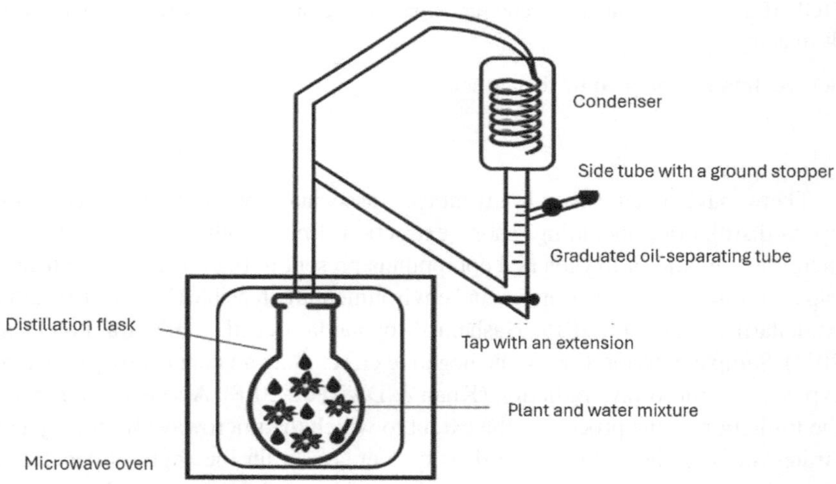

FIGURE 2.9 Diagrammatic representation of hybrid ohmic hydro-distillation method.

Source: Image generated by the author.

important factor in the design of ohmic-assisted hydro-distillation processes, as Gavahian et al. (2018) also noticed that metal ions were transferred from titanium electrodes into the extraction solution. An added disadvantage includes high capital costs in the construction of the ohmic process in comparison to microwave-assisted hydro-distillation (Gavahian & Farahnaky, 2018).

6. *Ultrasound-Assisted Hydro-Distillation* Ultrasound-assisted hydro-distillation is a complementary and future green extraction method that utilises increasing particle speed and movement through acoustic waves, causing the penetration and damage of the plant cells with the water solvent (Chen et al., 2021). The implosion of cavitation bubbles formed during ultrasonication produces micro-jets that disrupt essential oil glands, aiding the mass transfer and release of plant essential oil. The effectiveness of the cavitation phenomenon is dependent on operational parameters such as the frequency and intensity of the ultrasound rays, temperature, and extraction time (Li et al., 2014; Chen et al., 2021). There are two types of ultrasound-assisted hydro-distillation processes, namely ultrasound pre-treatment and simultaneous ultrasound and hydro-distillation. Processes incorporating ultrasound pre-treatment utilise an ultrasound probe in the water and plant material mixture before entering a conventional hydro-distillation Clevenger apparatus (Chen et al., 2021). Simultaneous ultrasound-assisted hydro-distillation incorporates an ultrasound probe into the traditional Clevenger hydro-distillation systems, as depicted in Figure 2.10.

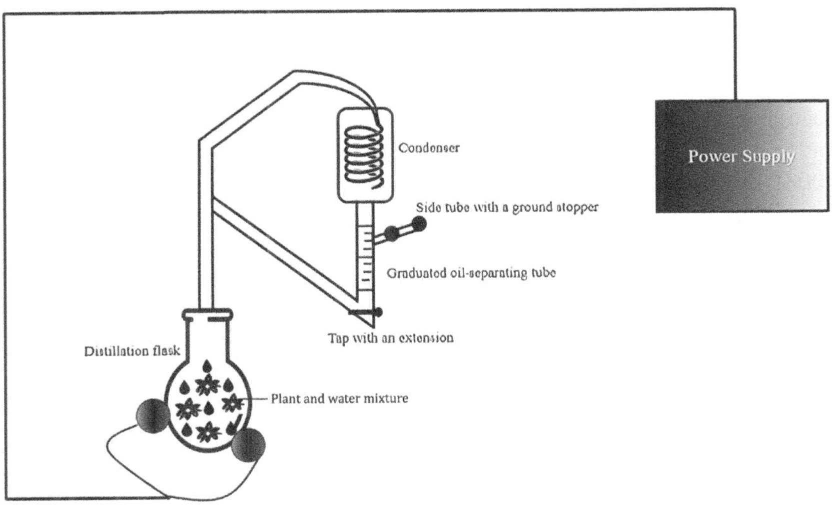

FIGURE 2.10 Illustration showcasing ultrasound-assisted hydro-distillation extraction process.

Image generated by the author.

The advantages of both forms of ultrasound-assisted hydro-distillation have shown accelerated extraction rates, quicker extraction times, reduced energy consumption, and less thermal degradation; hence, a higher quality and good flavour oil is produced (Chen et al., 2021; Li et al., 2014). The drawback of ultrasound-assisted hydro-distillation processes is that the metal contamination from the sonotrode in the mixture leads to increased oxidation and decreased essential oil stability (Pingret et al., 2013). Hence, the choice of sonotrode is essential to the design of the ultrasound-assisted hydro-distillation process (Li et al., 2014).

7. Microwave Hydro-Diffusion Recently, the demand for novel extraction techniques has been increasing. These techniques are required to be capable of automation, with reduced durations of extraction and decreased consumption of organic solvents. Additionally, these novel techniques are required to prevent pollution and minimise costs associated with sample preparation (Vian et al., 2008). Considering these goals, microwave extraction (see Figure 2.11) has advanced, resulting in various microwave-assisted extraction techniques – to the point that microwave-assisted extraction techniques have replaced some traditional extraction and thermal techniques (Vian et al., 2008).

The study of Vian et al. (2008) entailed the design and development of a green technique to extract essential oils using gravity coupled with microwave hydro-diffusion. The innovative process is considered a unique “inverse” microwave alembic, which couples gravity and microwave heating at atmospheric pressure. Microwave hydro-diffusion was developed for laboratory and commercial-scale applications entailing essential oil extraction from various aromatic plants. The method entails the insertion of plant material within a microwave reactor devoid of any water or solvent (Vian et al., 2008).

Due to the interior heating of the water contained within the plant material, the plant’s glands and oleiferous receptacles become ruptured. The microwave heating action plays a vital role in freeing essential oil and *in-situ* water, which undergo transfer from within to the outside of the vegetal material (Vian et al., 2008). In this process, hydro-diffusion enables the extract (comprising essential oil and water) to be diffused outside the vegetal material to gravity fall out of the microwave reactor through a pierced Pyrex disc. The extract undergoes continuous cooling using an external cooling system relative to the microwave. The extract undergoes collection and subsequent separation in a “Florentine flask”. Due to the essential oil being lighter than water, it resides above the water. Hence, simplistic separation is achieved (Vian et al., 2008). Vian et al. (2008) outlined the primary inherent advantage of the technique as its ability to achieve the extraction of essential oils devoid of energy-consuming processes, such as evaporation and distillation. Additionally, the technique can produce concentrated essential oil devoid of any contaminants, artefacts, or residual solvents (Vian et al., 2008). It was further remarked that combining microwave extraction with gravity and hydro-diffusion boasts other advantages in terms of selectivity and yield, improved extraction duration, improved composition of essential oil, and no adverse environmental effects (Vian et al., 2008).

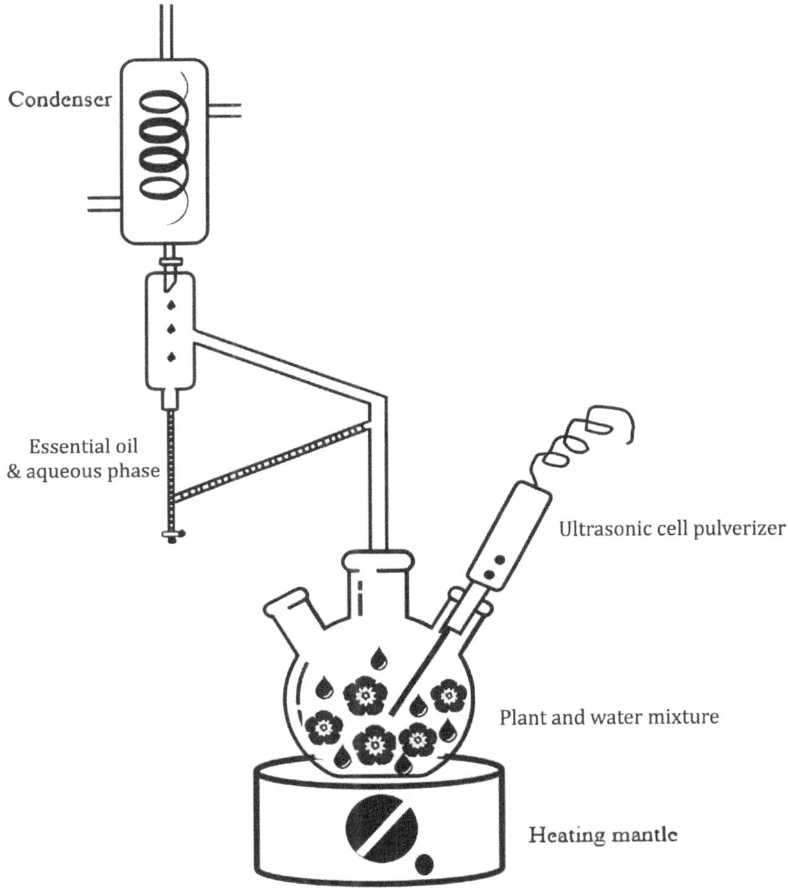

FIGURE 2.11 Graphical depiction of hybrid microwave hydro-diffusion process.

Source: Image generated by the author.

2.4 CASE STUDIES HIGHLIGHTING THE APPLICATION OF HYBRID TECHNIQUES

2.4.1 Efficiency, Yield, and Quality Considerations

Considering Table 2.1, the efficiency (time required) and yield of the essential oil extracted using hybrid techniques is significantly increased due to the synergistic effects of incorporating heating-assisted methods such as microwave, ohmic, or ultrasound to accelerate and improve the oil extraction process. Some studies have shown improved quality of essential oils (Hashemi et al., 2018), whilst others show no change in the quality of oil over traditional methods (Golmakani & Rezaei, 2008).

TABLE 2.1
Applications of hybrid techniques.

Extraction technique	Applications	References
Microwave-assisted steam distillation	Extraction of citrus oil from lemon, orange, and mandarin peels.	Shakir and Salih (2015)
	Separation of essential oil from tree peony (*Paeonia suffruticosa* Andrews) petals	Zhang et al. (2020)
	Essential oils extracted from two different herbs, namely basil and epazote	Cardoso-Ugarte et al. (2013)
Microwave-assisted water distillation	*Sideritis raeseri* essential oil extraction. The oil is commonly used in tea.	Drinić et al. (2021)
	Rosmarinus officinalis L. oil, which is used as a food flavouring agent, is extracted.	Jaimand et al. (2018)
	Extraction of essential oil from Navel Navelate oranges	Bustamante et al. (2016)
Ohmic-accelerated steam distillation	Essential oil extraction from lavender	Gavahian and Chu (2018)
Microwave-Assisted hydro-diffusion distillation	Essential oil extraction from *Acacia dealbata* flowers	López-Hortas et al. (2020)
	Essential oils from fresh ginger root (*Zingiber officinale* Roscoe)	Racoti et al. (2017)
Ohmic-assisted hydro-distillation	Ohmic heating-assisted hydro-distillation of clove essential oil	Tunç and Koca (2019)
	Ohmic-assisted hydro-distillation of essential oil from the aerial part of yarrow (*Achillea millefolium L.*)	Karami et al. (2022)
Ultrasound-assisted hydro-diffusion distillation	Essential oil extraction of Kalimantan white pepper oil	Gotama et al. (2023)
	Essential oil from *Iberis amara* seeds	Liu et al. (2019)

2.4.2 Environmental and Economic Implications

Overall, case studies have shown that hybrid techniques are environmentally safe and economically viable in comparison to traditional methods for solvent extraction. According to Solanki et al. (2018), ultrasound-assisted hydro-distillation has proven to be an alternative green extraction method, reflecting a 40% reduction in energy consumption and a noteworthy 47% decrease in carbon footprint. Gavahian et al. (2018) showed that ohmic-assisted hydro-distillation displayed 79% energy savings in comparison to hydro-distillation.

TABLE 2.2
Comparison of tradition versus hybrid techniques.

Extraction technique	Applications/comparative analysis	References
Microwave-assisted steam distillation	It is capable of achieving greater selectivities, reduced extraction durations, and improved extraction yields.	Chen et al. (2007)
	Efficient heat transfer, indirect contact with the source of heat, decreased extraction time, reduced energy requirements, and great duplicability.	Lukose et al. (2023)
	Oil extracted from dry lavender leaves flowers using microwave-assisted steam distillation and conventional steam distillation found the oils extracted were both qualitatively and quantitatively similar. However, the microwave steam distillation method occurred faster than the steam distillation method: 6 minutes compared to 30 minutes, respectively. It was found that microwave power is inversely proportional to the time needed to complete the extraction. However, it is essential to ensure the accurate control of temperature, as high temperatures promote the degradation of certain components within the oils.	Saldaña-Mendoza et al. (2022); Fardhyanti et al. (2019)
Microwave-assisted water distillation	Holds the potential to increase the quality of essential oils, shorten the extraction time frame, increase the process efficiency, and reduce the cost.	Akdağ and Öztürk (2019); Kaufmann and Christen (2002)
	The boiling point is reached more rapidly than in water distillation; this results in an increase in the efficiency of the extraction process.	Lucchesi et al. (2004); Flamini et al. (2007)
	Microwave-assisted water distillation has the ability to produce the same oil yield at a significantly reduced duration as compared to traditional water distillation; it has been approximated that time frames are 30 minutes and 4 hours, 30 minutes, respectively, for both methods to produce the same oil yield and quality.	Golmakani and Rezaei (2008)
Ohmic-accelerated steam distillation	A study investigated the effect of ohmic-accelerated steam distillation on the extraction of lavender oil from lavender inflorescences. The extraction time and chemical properties of the essential oil extracted were compared to those of traditional steam distillation. The ohmic-assisted method resulted in the same quantity of oil as steam distillation with a similar composition. It was also found that the relative densities and colours of the extracted essential oils were similar, and the ohmic-accelerated steam distillation did not adversely affect the valuable compounds of lavender essential oil.	Gavahian and Chu (2018)

(*Continued*)

TABLE 2.2 (*Continued*)
Comparison of tradition versus hybrid techniques.

Extraction technique	Applications/comparative analysis	References
Microwave-assisted hydro-diffusion distillation	A study was conducted to extract essential oil from oregano (*O. vulgare* L. spp. *hirtum*). The extraction time was reduced from 136 minutes to 24 minutes, with an increase in yield from 5.81% to 7.10% with microwave-assisted hydro-distillation. Additionally, the electrical consumption decreased from 1.360 kWh to 0.240 kWh and a reduced the CO_2 emission from 1.09 kg to 0.19 kg.	Drinić et al. (2020)
	The extraction of citrus essential oil was enhanced by the microwave process by depicting a reduced extraction time from 3 hours for hydro-distillation to 30 minutes for microwave-assisted. Additionally, it was shown that the yield improved from 0.21% for hydro-distillation to 0.24% for microwave-assisted and increased quantities of oxygenated compounds were found in the extracted oil.	Ferhat et al. (2016)
	A study investigated the extraction of essential oil from lemongrass (*Cymbopogon citratus*). It was shown that the time for extraction for microwave-assisted in comparison to hydro-distillation reduced from 180 minutes to 90 minutes, although with no significant difference in the constituents in the essential oil between the two methods.	Ranitha et al. (2014)
Ohmic-assisted hydro-distillation	Oil extraction of peppermint (*Mentha piperta*) by comparing microwave-assisted hydro-distillation, ohmic-assisted hydro-distillation, and conventional hydro-distillation. The study showed that microwave-assisted hydro-distillation and ohmic-assisted hydro-distillation collected the majority of the oil within 30 minutes, whilst hydro-distillation required 1 hour. The composition of the essential oil obtained from all methods produces a similar composition. Ohmic-assisted hydro-distillation demonstrated the most environmentally feasible extraction method presented in the study as it required the least amount of energy.	Gavahian et al. (2015)
	The extraction of Citronella oil by comparison with traditional hydro-distillation and ohmic-assisted hydro-distillation. The oil yield for both processes was constant at 1.5% w/w with similar physical properties. The advantages of the ohmic-assisted hydro-distillation process were a higher value oil (stronger antioxidant properties), 46% process time savings, and 79% energy savings in comparison to hydro-distillation.	Gavahian et al. (2018)

	Microwave-assisted hydro-distillation (2.5 minutes) and ohmic-assisted hydro-distillation (17.33 minutes) required a considerably lower time to initiate the essential oil production in contrast to hydro-distillation (40 minutes). However, it should be noted that microwave-assisted hydro-distillation had the fastest extraction time.	Manouchehri et al. (2018).
Ultrasound-assisted hydro-distillation	A study investigated the enhanced extraction ability of Palau leaves using an ultrasound probe prior to the traditional hydro-distillation process as a form of ultrasound pre-treatment. It was shown that the extraction time decreased, achieving higher extraction yield of essential oil, enhanced antioxidant and antibacterial activities, and no detectable quantities of metals were measured.	Hashemi et al. (2018)
	The combined process of the ultrasound-assisted and hydro-distillation processes using an ultrasound probe to extract oil from Iberis amara seeds. The study showed an increased yield of oil at 0.42% (v/w) in comparison with traditional hydro-distillation at 0.26% (v/w).	Liu et al. (2019)
	A simultaneous ultrasound-assisted and hydro-distillation process was determined to extract oil from java citronella grass. The optimum yield of oil was calculated to be 4.118% (w/w) at 21 minutes of extraction time compared to hydro-distillation with only a 2.33% (w/w) yield of oil. Ultrasound-assisted hydro-distillation has proven to be an alternative green extraction method, reflecting a 40% reduction in energy consumption and a noteworthy 47% decrease in carbon footprint.	Solanki et al. (2018)

2.5 FUTURE PROSPECTS AND INNOVATIONS

2.5.1 Ongoing Research and Development in Azeotropic Distillation

According to Aquilon et al. (2023), the commonly used methods in azeotropic distillation include extractive and pressure-swing distillation. However, due to the energy intensity of these processes, hybrid distillation is recommended. As such, the ongoing research and development in azeotropic distillation include the selection and optimisation of appropriate solvents to formulate azeotropic mixtures with enhanced properties for essential oil extraction, including considerations for solvent efficiency, environmental impact, and ease of recovery (Aquilon et al., 2023). Optimised distillation configurations can also be considered to improve the efficiency of the azeotropic distillation process, where the interest lies in optimising parameters such as temperature, pressure, and extraction time to enhance overall efficiency. Additionally, the commercial scale-up of such processes is an area where advancements may be sought. Advanced separation techniques involving the use of membranes and adsorption (among others) may also attract consideration to improve the separation of essential oils from azeotropic mixtures (Aquilon et al., 2023). For improved precision and reproducibility in azeotropic distillation processes, advances in automation and control systems can be considered to optimise essential oil extraction conditions by involving sensors and real-time monitoring (Aquilon et al., 2023). Due to the increasing focus on sustainability, developing environmentally friendly solvents, reducing wastes, and improving energy efficiency in azeotropic distillation are anticipated to gain consideration. In the same vein, research may focus on improving azeotropic distillation's health and safety aspects, addressing concerns regarding solvent toxicity, and ensuring that the final essential oil product meets quality and safety standards.

2.5.2 Potential Advancements and Improvements

Although Sun et al. (2019) recommend that the price and source of solvents be carefully screened, a further potential advancement in azeotropic distillation for essential oil extraction entails formulating environmentally friendly or "green" solvents for azeotropic distillation. Such solvents are characterised as being sustainable, non-toxic, and having minimal environmental impact. Nanomaterials can also be considered a process intensification strategy due to their potential to offer improved surface area and adsorption capabilities, contributing to better extraction yields. According to Aquilon et al. (2023), intensification and integration techniques are used to reduce capital costs and energy consumption in azeotropic distillation systems but are more complex than traditional processes. As specialised configurations are created, a comparison of specialised and traditional configurations should be made regarding their financial costs and environmental performances. Other process intensification strategies to improve the efficiency of azeotropic distillation could include the use of advanced heat exchangers, novel reactor designs, and improved process control strategies.

Within these considerations lies the development of methods that facilitates the selective extraction of specific compounds from plant materials, which could fine-tune the azeotropic mixture to target constituents in essential oils. Despite the perceived benefits of process intensification, Ma et al. (2019) showed that, even though some process intensification technologies increase economic efficiency and reduce energy consumption, they also make the process more complex to operate and monitor. Hence, process intensification strategies must be thoroughly vetted prior to implementation.

Additionally, the design of advanced recovery systems for azeotropic solvents is a potential improvement to minimise waste and improve the sustainability of the process. Advanced sensors, analytics, automation, and robotics can also be employed for real-time monitoring, enhanced precision, and repeatability to optimise process parameters, ensuring consistent and high-quality essential oil production. However, in this regard, Aquilon et al. (2023) advise that the lack of established design standards for distillation is challenging for researchers and industry practitioners. Therefore, establishing thorough standards that consider varied reaction systems, feed compositions, and operating circumstances is vital to maintain consistency and dependability in distillation process design and scale-up. Furthermore, combining azeotropic distillation with other extraction methods to create hybrid processes that offer synergistic advantages and improved extraction yields is a potential advancement. However, Zhang et al. (2018) reported that optimisation studies should consider multiple parameters to ensure that desired outcomes are reached. Finally, the development of enhanced safety protocols to address potential risks associated with the use of solvents in azeotropic distillation, including innovations in containment systems and emergency response mechanisms and research on effective scale-up strategies for azeotropic distillation processes to facilitate their application in large-scale industrial settings are potential improvements.

2.5.3 The Role of Azeotropic Distillation in the Future of Essential Oil Extraction

When considering the role of azeotropic distillation in the future of essential oil extraction, developing environmentally friendly solvents can help address sustainability and environmental concerns associated with essential oil extraction. This notion is supported by Aquilon et al. (2023), who believe that as the global chemical industry shifts towards sustainability, the integration of renewable feedstocks is anticipated to become more important in future. In the same vein, Guo and Wang (2019) recommend that the most advanced technology and materials should be applied in azeotropic distillation processes, as the choice of azeotropic agent greatly influences the azeotropic distillation process. Such transitions towards sustainable feedstocks are consistent with the industry's overall green and circular chemistry trend. Furthermore, developing techniques for converting waste streams into valuable products can contribute to the circular economy.

According to Aquilon et al. (2023), technological breakthroughs such as process control and automation are two important areas of growth. Artificial intelligence and machine learning algorithms in advanced control systems can optimise operations in real time to improve process efficiency and increase responsiveness to changing feedstock compositions. Additionally, online sensors and advanced analytics will be critical in monitoring and improving operations. These technologies allow for real-time data analysis, enabling faster decision-making and proactive modifications to maintain optimal performance and product quality. On the other hand, ongoing research regarding novel solvents and azeotropic mixtures may result in improved extraction efficiency, reduced environmental impact, and enhanced safety for essential oil production. It is anticipated that such advancements in azeotropic distillation may lead to more selective extraction processes, enabling the targeted isolation of compounds from plant materials, which is valuable for industries where the composition of essential oils is crucial.

It is also anticipated that combining azeotropic distillation with other extraction processes (hybrid approaches) could become more common, offering greater benefits in extracting a broader range of compounds from diverse plant sources. According to Aquilon et al. (2023), integrating distillation with other novel technologies offers a potential development horizon, as hybrid processes that combine distillation with separation technologies improve overall process efficiency and minimise energy usage. As such, customising azeotropic distillation processes to suit the characteristics of different plant sources could become a standard practice to optimise extraction efficiency and product quality. Such hybrid approaches are supported by Guo and Wang (2019), who believe that it is important to find new separation methods and to reform existing processes to improve the economic and environmental benefits of traditional azeotropic distillation processes. Hence, they regard the reduction of energy consumption as one of the important directions of azeotropic distillation technology development.

Finally, considering that scale-up strategies for azeotropic distillation processes can aid in meeting the commercial demands of the essential oil industry, continued emphasis on developing safety protocols and ensuring compliance with regulatory standards remain crucial for the adoption of azeotropic distillation methods in the essential oil industry.

2.6 CONCLUSION

Azeotropic distillation processes such as steam distillation, steam-water distillation, hydro-diffusion distillation, and hydro-distillation are conventional extraction methods used for centuries to extract oils from various organic plant materials. The low yield and long extraction times have sparked new research into developing new green extraction methods reviewed in this chapter, including, but limited to, microwave-assisted, ohmic-assisted, solar-assisted, and ultrasound-assisted azeotropic distillation systems. All these techniques facilitate the conventional process by enhancing the internal heating capacity of the plant material, thereby promoting the efficient release of essential oil from plant cells. These methods offer economic and environmental viability through energy conservation, lowered carbon emissions, increased yields, and reduced extraction time. They emerge as effective alternatives to traditional azeotropic distillation, shaping the future of essential oil extraction.

This chapter carefully outlined the limitations and shortcomings of traditional azeotropic distillation with a view of indicating the potential improvements that can be made to improve the technology towards commercial applications for essential oil extraction. Through this lens, it is anticipated that azeotropic distillation for essential oil extraction will endure well into the future. With the advancement of technologies, such as real-time monitoring and control, the implementation of advanced analytics and safety features, the view is that azeotropic distillation will become more effective over time to ensure efficient and quality essential oil extraction.

REFERENCES

Abdellatif, F., & Hassani, A. (2015). Chemical composition of the essential oils from leaves of Melissa officinalis extracted by hydro-distillation, steam distillation, organic solvent and microwave hydro-distillation. *Journal of Materials and Environmental Science*, *6*(1), 207–213.

Abert Vian, M., Fernandez, X., Visinoni, F., & Chemat, F. (2008). Microwave hydrodiffusion and gravity, a new technique for extraction of essential oils. *Journal of Chromatography A*, *1190*(1–2), 14–17.

Akdağ, A., & Öztürk, E. (2019). Distillation methods of essential oils. *Selçuk Üniversitesi Fen Fakültesi Fen Dergisi*, *45*(1), 22–31.

Aquilon, A. F., Cargullo, D. M., Onayan, J., Sarno, J., Molino, V. M., & Lopez, E. C. R. (2023). Recent advances in extractive distillation. *Engineering Proceedings*, *56*(1), 11.

Battaglia, S. (2003). *The complete guide to aromatherapy* (pp. 1–602). Queensland, AU: International Centre of Holistic Aromatherapy.

Bustamante, J., van Stempvoort, S., García-Gallarreta, M., Houghton, J. A., Briers, H. K., Budarin, V. L., & Clark, J. H. (2016). Microwave-assisted hydro-distillation of essential oils from wet citrus peel waste. *Journal of Cleaner Production*, *137*, 598–605.

Cardoso-Ugarte, G. A., Juárez-Becerra, G. P., SosaMorales, M. E., & López-Malo, A. (2013). Microwave-assisted extraction of essential oils from herbs. *Journal of Microwave Power and Electromagnetic Energy*, *47*(1), 63–72.

Chen, F., Sun, Y., Zhao, G., Liao, X., Hu, X., Wu, J., & Wang, Z. (2007). Optimization of ultrasound-assisted extraction of anthocyanins in red raspberries and identification of anthocyanins in extract using high-performance liquid chromatography–mass spectrometry. *Ultrasonics Sonochemistry*, *14*(6), 767–778.

Chen, G., Sun, F., Wang, S., Wang, W., Dong, J., & Gao, F. (2021). Enhanced extraction of essential oil from Cinnamomum cassia bark by ultrasound assisted hydrodistillation. *Chinese Journal of Chemical Engineering*, *36*, 38–46.

Chouhan, K. B. S., Tandey, R., Sen, K. K., Mehta, R., & Mandal, V. (2019). Critical analysis of microwave hydrodiffusion and gravity as a green tool for extraction of essential oils: Time to replace traditional distillation. *Trends in Food Science & Technology*, *92*, 12–21.

Craveiro, A. A., Matos, F. J. A., Alencar, J. W., & Plumel, M. M. (1989). Microwave oven extraction of an essential oil. *Flavour and Fragrance Journal*, *4*(1), 43–44.

Di Giacomo, A., & Di Giacomo, G. (2002). Essential oil production. In *Citrus: The genus citrus* (pp. 114–147). CRC Press.

Distillation, s.v. "Cambridge Dictionary", last modified 8 August 2023, https://dictionary.cambridge.org/dictionary/english/distillation

Drinić, Z., Pljevljakušić, D., Janković, T., Zdunić, G., Bigović, D., & Šavikin, K. (2021). Hydro-distillation and microwave-assisted distillation of Sideritis raeseri: Comparison of the composition of the essential oil, hydrolat and residual water extract. *Sustainable Chemistry and Pharmacy*, *24*, 100538.

Drinić, Z., Pljevljakušić, D., Živković, J., Bigović, D., & Šavikin, K. (2020). Microwave-assisted extraction of O. vulgare L. spp. hirtum essential oil: Comparison with conventional hydro-distillation. *Food and Bioproducts Processing*, *120*, 158–165.

Fardhyanti, D. S., Sediawan, W. B., & Hisyam, A. (2019). Kinetics of mace (Myristicae arillus) essential oil extraction using microwave assisted hydrodistillation: Effect of microwave power. *Industrial Crops and Products*, *131*, 315–322.

Ferhat, M. A., Boukhatem, M. N., Hazzit, M., Meklati, B. Y., & Chemat, F. (2016). Cold pressing, hydrodistillation and microwave dry distillation of citrus essential oil from Algeria: A comparative study. *Electronic Journal of Biology S*, *1*, 30–41.

Ferhat, M. A., Meklati, B. Y., Smadja, J., & Chemat, F. (2006). An improved microwave Clevenger apparatus for distillation of essential oils from orange peel. *Journal of Chromatography A*, *1112*(1–2), 121–126.

Flamini, G., Tebano, M., Cioni, P. L., Ceccarini, L., Ricci, A. S., & Longo, I. (2007). Comparison between the conventional method of extraction of essential oil of Laurus nobilis L. and a novel method which uses microwaves applied in situ, without resorting to an oven. *Journal of Chromatography A*, *1143*(1–2), 36–40.

Gavahian, M., & Chu, Y. H. (2018). Ohmic accelerated steam distillation of essential oil from lavender in comparison with conventional steam distillation. *Innovative Food Science & Emerging Technologies*, *50*, 34–41.

Gavahian, M., & Farahnaky, A. (2018). Ohmic-assisted hydrodistillation technology: A review. *Trends in Food Science & Technology*, *72*, 153–161.

Gavahian, M., Farahnaky, A., Farhoosh, R., Javidnia, K., & Shahidi, F. (2015). Extraction of essential oils from Mentha piperita using advanced techniques: Microwave versus ohmic assisted hydrodistillation. *Food and Bioproducts Processing*, *94*, 50–58.

Gavahian, M., Farahnaky, A., Javidnia, K., & Majzoobi, M. (2012). Comparison of ohmic-assisted hydrodistillation with traditional hydrodistillation for the extraction of essential oils from Thymus vulgaris L. *Innovative Food Science & Emerging Technologies*, *14*, 85–91.

Gavahian, M., Lee, Y. T., & Chu, Y. H. (2018). Ohmic-assisted hydrodistillation of citronella oil from Taiwanese citronella grass: Impacts on the essential oil and extraction medium. *Innovative Food Science & Emerging Technologies*, *48*, 33–41.

Golmakani, M. T., & Rezaei, K. (2008). Comparison of microwave-assisted hydrodistillation withthe traditional hydrodistillation method in the extractionof essential oils from Thymus vulgaris L. *Food Chemistry*, *109*(4), 925–930.

Gotama, B., Ginting, R. R., Pusfitasari, M. D., & Wibowo, A. A. (2023, September). Evaluation of the kinetics of microwave-assisted hydro distillation (MAHD) and ultrasound-followed hydro distillation (US-HD) for the extraction of Kalimantan white pepper oil. In *AIP Conference Proceedings* (Vol. 2933, No. 1). AIP Publishing.

Guo, Y., & Wang, L. (2019). Research progress on azeotropic distillation technology. *Advances in Chemical Engineering and Science*, *9*(04), 333.

Hashemi, S. M. B., Khaneghah, A. M., Koubaa, M., Barba, F. J., Abedi, E., Niakousari, M., & Tavakoli, J. (2018). Extraction of essential oil from Aloysia citriodora Palau leaves using continuous and pulsed ultrasound: Kinetics, antioxidant activity and antimicrobial properties. *Process Biochemistry*, *65*, 197–204.

Hashemi-Moghaddam, H., Mohammadhosseini, M., & Azizi, Z. (2018). Impact of amine- and phenyl-functionalized magnetic nanoparticles impacts on microwave-assisted extraction of essential oils from root of Berberis integerrima Bunge. *Journal of Applied Research on Medicinal and Aromatic Plants*, *10*, 1–8.

Ibrahim, M., Ankwai, G. E., Gungshik, J. R., & Taave, P. (2021). Comparative extraction of essential oils of Mentha piperita (mint) by steam distillation and enfleurage. *Nigerian Journal of Chemical Research*, *26*(2), 56–62.

Jaimand, K., Bagher Rezaee, M., & Homami, S. (2018). Comparison extraction methods of essential oils of Rosmarinus officinalis L. In Iran by microwave assisted water distillation; water distillation and steam distillation. *Journal of Medicinal Plants and By-Product*, *7*(1), 9–14.

Kabuba, J., & Huberts, R. (2009). Steam extraction of essential oils: Investigation of process parameters. *The Canadian Journal of Chemical Engineering*, *87*(6), 915–920.

Kant, R., & Kumar, A. (2021). Advancements in steam distillation system for oil extraction from peppermint leaves. *Materials Today: Proceedings*, *47*, 5794–5799.

Kant, R., & Kumar, A. (2022). Review on essential oil extraction from aromatic and medicinal plants: Techniques, performance and economic analysis. *Sustainable Chemistry and Pharmacy*, *30*, 100829.

Karami, P., Zandi, M., & Ganjloo, A. (2022). Evaluation of key parameters during ohmic-assisted hydro-distillation of essential oil from aerial part of yarrow (Achillea millefolium L.). *Journal of Applied Research on Medicinal and Aromatic Plants*, *31*, 100425.

Kaufmann, B., & Christen, P. (2002). Recent extraction techniques for natural products: Microwave-assisted extraction and pressurised solvent extraction. *Phytochemical Analysis: An International Journal of Plant Chemical and Biochemical Techniques*, *13*(2), 105–113.

Khalisanni, K., Ahmad, A. W., Faridatul, A. M. N., & Bidin, I. (2014). Enhancing efficiency at the distillation process by using estimated virtual exhaustive (EVE) technique. *Innovative Plant Productivity and Quality*, *129*.

Khan, M. F., & Dwivedi, A. K. (2018). A review on techniques available for the extraction of essential oils from various plants. *International Research Journal of Engineering and Technology*, *5*(5), 5–8.

Koller, J., Baumer, U., Kaup, Y., & Weser, U. (2005). Herodotus' and Pliny's embalming materials identified on ancient Egyptian mummies. *Archaeometry*, *47*(3), 609–628.

Li, Y., Fabiano-Tixier, A. S., & Chemat, F. (2014). *Essential oils as reagents in green chemistry* (Vol. 1, pp. 71–78). Cham, Switzerland: Springer International Publishing.

Liu, X. Y., Ou, H., Xiang, Z. B., & Gregersen, H. (2019). Optimization, chemical constituents and bioactivity of essential oil from Iberis amara seeds extracted by ultrasound-assisted hydro-distillation compared to conventional techniques. *Journal of Applied Research on Medicinal and Aromatic Plants*, *13*, 100204.

López-Hortas, L., Falqué, E., Domínguez, H., & Torres, M. D. (2020). Microwave hydrodiffusion and gravity versus conventional distillation for Acacia dealbata flowers. Recovery of bioactive extracts for cosmetic purposes. *Journal of Cleaner Production*, *274*, 123143.

Lucchesi, M. E., Chemat, F., & Smadja, J. (2004). Solvent-free microwave extraction of essential oil from aromatic herbs: Comparison with conventional hydro-distillation. *Journal of Chromatography A*, *1043*(2), 323–327.

Lucchesi, M. E., Smadja, J., Bradshaw, S., Louw, W., & Chemat, F. (2007). Solvent free microwave extraction of Elletaria cardamomum L.: A multivariate study of a new technique for the extraction of essential oil. *Journal of Food Engineering*, *79*(3), 1079–1086.

Lukose, R., Kalbande, S. R., & John, J. (2023). Utilisation of solar energy for essential oil extraction: An overview. *International Journal of Ambient Energy*, *44*(1), 1151–1165.

Ma, K., Yu, M., Dai, Y., Ma, Y., Gao, J., Cui, P., & Wang, Y. (2019). Control of an energy-saving side-stream extractive distillation process with different disturbance conditions. *Separation and Purification Technology*, *210*, 195–208.

Manouchehri, R., Saharkhiz, M. J., Karami, A., & Niakousari, M. (2018). Extraction of essential oils from damask rose using green and conventional techniques: Microwave and ohmic assisted hydrodistillation versus hydrodistillation. *Sustainable Chemistry and Pharmacy*, *8*, 76–81.

Manzan, A. C. C., Toniolo, F. S., Bredow, E., & Povh, N. P. (2003). Extraction of essential oil and pigments from Curcuma longa [L.] by steam distillation and extraction with volatile solvents. *Journal of Agricultural and Food Chemistry*, *51*(23), 6802–6807.

Marković, M. S., Bošković-Vragolović, N., Ristić, M. S., Pavićević, V., Veljković, V. B., & Milojević, S. (2017). Fractionation of the essential oil from juniper (Juniperus et al.) berries by hydrodistillation and rectification. *Hemijska industrija*, *71*(6), 471–477.

Masango, P. (2005). Cleaner production of essential oils by steam distillation. *Journal of Cleaner Production*, *13*(8), 833–839.

Masum, Z., Mahfud, M., & Altway, A. (2019). Parameter for scale-up of extraction Cymbopogon nardus dry leaf using microwave assisted hydro-distillation. *Journal of Applied Engineering Science*, *17*(2).

Nannaware, A. D., Kumar, C. M. S., Srivastava, S., Singh, S., Gupta, M. K., Rout, P. K., . . . & Roy, S. (2022). Eco-friendly solar distillation apparatus for improving the yield of essential oils with enhancing organoleptic characteristics. *Renewable Energy*, *191*, 345–356.

Nourafcan, H., FatemehSefidkon, A. K., Mousavi, A., & Sharifi, M. (2014). Effects of different distillation methods on essential oil content and composition of Lippiacitriodora HBK. *Journal of Biodiversity and Environmental Sciences (JBES)*, *5*(3), 115–123.

O'Hara, G. (1998). The magic of aromatherapy. *Llewellyn Publications, St. Paul, Minnesota*, *7*, 31.

Öztekin, S., & Soysal, Y. (1998). Tibbi Ve Aromatik Bitkilerde Ekstraksiyon Yöntemleri. Tarımsal Mekanizasyon 18, Ulusal Kongresi, Tekirdağ (731–745.ss.).

Parikh, J. K., & Desai, M. A. (2011). Hydrodistillation of essential oil from Cymbopogon flexuosus. *International Journal of Food Engineering*, *7*(1).

Pingret, D., Fabiano-Tixier, A. S., & Chemat, F. (2013). Degradation during application of ultrasound in food processing: A review. *Food Control*, *31*(2), 593–606.

Price, S. (2000). *Aromatherapy workbook: A complete guide to understanding and using essential oils*. Harper Collins.

Rackley, S. A. (2017). *Carbon capture and storage*. Butterworth-Heinemann.

Racoti, A., Buttress, A. J., Binner, E., Dodds, C., Trifan, A., & Calinescu, I. (2017). Microwave assisted hydro-distillation of essential oils from fresh ginger root (*Zingiber officinale* Roscoe). *Journal of Essential Oil Research*, *29*(6), 471–480.

Radwan, M. N., Morad, M. M., Ali, M. M., & Wasfy, K. I. (2020). A solar steam distillation system for extracting lavender volatile oil. *Energy Reports*, *6*, 3080–3087.

Ranitha, M., Nour, A. H., Sulaiman, Z. A., & Nour, A. H. (2014). A comparative study of lemongrass (Cymbopogon Citratus) essential oil extracted by microwave-assisted hydrodistillation (MAHD) and conventional hydrodistillation (HD) method. *International Journal of Chemical Engineering and Applications*, *5*(2), 104.

Rasul, M. G. (2018). Conventional extraction methods use in medicinal plants, their advantages and disadvantages. *International Journal of Basic Sciences and Applied Computing*, *2*, 10–14.

Rezvanpanah, S., Rezaei, K. Razavi, S. H., & Moini, S. (2008). Use of microwave-assisted hydrodistillation to extract the essential oils from Satureja hortensis and Satureja montana. *Food Science and Technology Research*, *14*(3), 311–314.

Richardson, J. F. (2002). *Coulson and Richardson's chemical engineering*. Elsevier Science.

Saldaña-Mendoza, S. A., Chávez-González, M. L., Ramírez-Guzmán, N., Pacios-Michelena, S., & Aguilar, C. N. (2022). Technological trends in the extraction of essential oils. *Environmental Quality Management*, *32*(1), 441–450.

Izquierdo, S. S., Bermello Crespo, A., Márquez Conde, T., López Hernández, O. D., González Sanabia, M. L., & Llópiz Yurell, J. C. (2013). Extracción asistida por microondas de lípidos de las semillas de Cucurbita pepo L.(calabaza). *Revista cubana de plantas medicinales*, *18*(1), 17–26.

Salonen, L., Kaihola, L., Carter, B., Cook, G. T., & Passo Jr, C. J. (2012). Environmental liquid scintillation analysis. In *Handbook of radioactivity analysis* (pp. 625–693). Academic Press.

Schlosser, Š. (2011, May). Distillation–from Bronze Age till today. In *38th International Conference of Slovak Society of Chemical Engineering, Tatranské Matliare, Slovakia.*

Seedat, N., Kauchali, S., & Patel, B. (2020). Novel representation of vapour-liquid equilibrium curves for multicomponent systems: Design of total reflux distillation columns. *Chemical Engineering Research and Design*, *155*, 22–39.

Shakir, I. K., & Salih, S. J. (2015). Extraction of essential oils from citrus by-products using microwave steam distillation. *Iraqi Journal of Chemical and Petroleum Engineering*, *16*(3), 11–22.

Solanki, K. P., Desai, M. A., & Parikh, J. K. (2018). Sono hydrodistillation for isolation of citronella oil: A symbiotic effect of sonication and hydrodistillation towards energy efficiency and environment friendliness. *Ultrasonics Sonochemistry*, *49*, 145–153.

Stratakos, A. C., & Koidis, A. (2016). Methods for extracting essential oils. In *Essential oils in food preservation, flavor and safety* (pp. 31–38). Academic Press.

Sun, S., Lü, L., Yang, A., & Shen, W. (2019). Extractive distillation: Advances in conceptual design, solvent selection, and separation strategies. *Chinese Journal of Chemical Engineering*, *27*(6), 1247–1256.

Tunç, M. T., & Koca, I. (2019). Ohmic heating assisted hydrodistillation of clove essential oil. *Industrial Crops and Products*, *141*, 111763.

Van Winkle, M. (1967). *Distillation*. United States of America: McGraw-Hill.

Yusoff, Z. M., Muhammad, Z., Kasuan, N., Rahiman, M. H. F., & Taib, M. N. (2013). Effect of temperature on kaffir lime oil by using hydro-diffusion steam distillation system. *The Malaysian Journal of Analytical Sciences*, *17*(2), 326–339.

Zhang, Q., Huo, R., Ma, Y. Yan, S., Yang, L., & Chen, F. (2020). A novel microwave-assisted steam distillation approach for separation of essential oil from tree peony (Paeonia suffruticosa Andrews) petals: Optimization, kinetic, chemical composition and antioxidant activity. *Industrial Crops and Products*, *154*, 112669.

Zhang, X., Li, X., Li, G., Zhu, Z., Wang, Y., & Xu, D. (2018). Determination of an optimum entrainer for extractive distillation based on an isovolatility curve at different pressures. *Separation and Purification Technology*, *201*, 79–95.

3 Soxhlet Solvent Extraction of Phytoconstituents

Samuel Olugbenga Olunusi and Nor Hanuni Ramli

3.1 INTRODUCTION

High-quality essential oil production relies on a comprehensive and knowledge-intensive extraction approach. It is crucial to recognize that the selection of an extraction method for retrieving essential oils and other volatile substances from plant components significantly impacts chemical composition, extract yield, and the overall quality of the raw product, thus influencing the viability of the entire operation. While various techniques exist for extracting volatile constituents from plant parts, authentic essential oils are obtained exclusively through distillation methods. Distillation, involving the physical separation of mixtures based on differences in volatilities within a boiling liquid mixture, is the primary method. However, due to the heat sensitivity of certain bioactive compounds, the distillation method may be impractical, leading to the adoption of solvent extraction (Hunter, 2009).

3.1.1 Background and Significance

Solvent extraction stands out as a widely employed method for obtaining essential oils from plants, involving the batch or continuous mass transfer of the desired compound into a polar or nonpolar extraction solvent (Ciuca & Racovita, 2023). This method offers a selective approach for extracting bioactive compounds, determined by the chosen solvent, influencing the product yield. Effective recovery of analytes or bioactive compounds during solvent extraction relies on two crucial analytical steps: solvent selection and solid sample preparation. Solid preparation facilitates solvent percolation through the matrix, while solvent selection depends on plant type, specific plant part, bioactive compound nature, and solvent availability. Generally, nonpolar solvents like hexane and dichloromethane are used for nonpolar compounds, and polar solvents such as water, methanol, and ethanol are used for polar compounds (Abubakar & Haque, 2020). Common practice involves choosing two miscible solvents like water/dichloromethane or water/hexane for

DOI: 10.1201/9781032634180-3

liquid–liquid extraction. The presence of water in the solvent is crucial due to its high polarity, miscibility with organic solvents, and aiding in ease of separation after extraction. Successful yield efficiency is achieved when solvents and the intended solute exhibit similar polarity (Kapadia et al., 2022). Selected solvents typically possess low boiling points, are nontoxic, nonflammable, and chemically neutral, contributing to the stability of the extracted compound (Arceusz et al., 2013). Solvent extraction's significance lies in its applicability to delicate plants and flowers susceptible to heat degradation, the extraction of valuable substances, and solvent purification by removing undesirable components.

3.2 OBJECTIVES OF THE CHAPTER

This chapter aims to describe the past, the present, and the panacea of Soxhlet extraction (SE) under the following objectives:

1. To systematically describe Soxhlet Extractor, its fundamental theory, and operating principle.
2. To describe its yield enhancement and optimization techniques with case studies.
3. To analyze the environmental impact and sustainability of SE method.
4. To explore its limitations and future prospects in extracting essential oils.

3.3 OVERVIEW OF ESSENTIAL OILS

Essential oils represent volatile, complex secondary metabolites synthesized by oil-yielding plants, possessing distinctive characteristics such as odor, taste, diverse functional groups, and repellent activity, often attributed to the presence of monoterpenes and sesquiterpenes derived from isoprene (Nerio et al., 2010). Typically consisting of 20 or more bioactive substances per essential oil, they are generally acknowledged as safe for extraction methods include azeotropic distillation techniques like hydro-distillation, hydro-diffusion, and steam distillation (SD), as well as solvent extraction methods.

Essential oils exhibit organic, highly concentrated, and volatile properties, being soluble in alcohol, mostly colorless, hydrophobic, less viscous than oil, less dense than water, possessing a watery texture, and characterized by a highly aromatic nature (Aier & Panwar, 2022). Beyond their aromatic features, essential oils showcase numerous antimicrobial and antioxidant properties, rendering them promising bioactive additions to various applications, including edible food coatings, perfumery, dentistry, food flavoring, preservatives, sanitary and cosmetic products, healing therapies, cleaning product fragrances, and industrial solvents (Jugreet & Mahomoodally, 2020). Additionally, they exhibit antifungal, insecticidal, and repellent activities. Examples of aromatic plant parts containing essential oils encompass flowers (e.g., orange, pink, lavender), leaves (e.g., eucalyptus, mint, thyme, bay leaf, savory, sage, pine needles), tree underground organs (e.g., vetiver), rhizomes (e.g., ginger, sweet flag), seeds (e.g., carvi, coriander), fruits (e.g., fennel, anise, citrus epicarps), and wood and bark (e.g., cinnamon, sandalwood, rosewood) (Dhifi et al., 2016).

3.4 BRIEF HISTORY OF SE

In 1879, the Soxhlet extractor was conceptualized by Franz Ritter Von Soxhlet, a German agricultural chemist, renowned for its primary application in extracting fat from milk solids. This innovative design incorporated a constant-level siphon, facilitating the return of the extract to the solvent flask upon completing an extraction cycle. Over the subsequent decades, the Soxhlet extractor underwent evolutionary phases, marked by terms like displacement, maceration, decoction, infusion, and lixiviation in the mid-nineteenth and twentieth centuries. Approximately 30 extractors emerged during this period, with a quarter of them representing variations of the original Soxhlet design, named in homage to its creator. Soxhlet's motivation for this innovation stemmed from the quest to determine quantitatively the fat content of organic matter. Historically, solid–liquid extraction practices, integral to processes like tea and perfume preparation, preceded the advent of Soxhlet's invention (Jensen, 2007).

3.5 PRINCIPLES OF SE

3.5.1 Theory of SE

SE operates on the principle of reflux (Figure 3.1), which involves the condensation of vapors with the condensate returning to the source, forming a continuous cycle. This method, known for its boiling and condensation theory, represents a liquid–solid extraction process conducted at a constant temperature. During the specified time frame, the solvent is recycled through evaporation and condensation without any loss. The Soxhlet extractor is a specific apparatus employed in this process. This type of liquid–solid extraction is particularly suitable when the desired chemical compounds have partial solubility in the chosen solvent, and the impurities (residual materials post-extraction) remain insoluble. Predominantly used in herbal industries, Soxhlet solvent extraction proves efficient, user-friendly, and cost-effective (D.-G. Wang et al., 2013).

3.5.2 Working Principle

The Soxhlet extractor consists of three parts that are fixed systematically to operate refluxing and siphoning principle automatically. These parts are:

1. *Round bottom flask* meant to boil the solvent and thereafter houses the mixture of the solvent and the bioactive or chemical compounds.

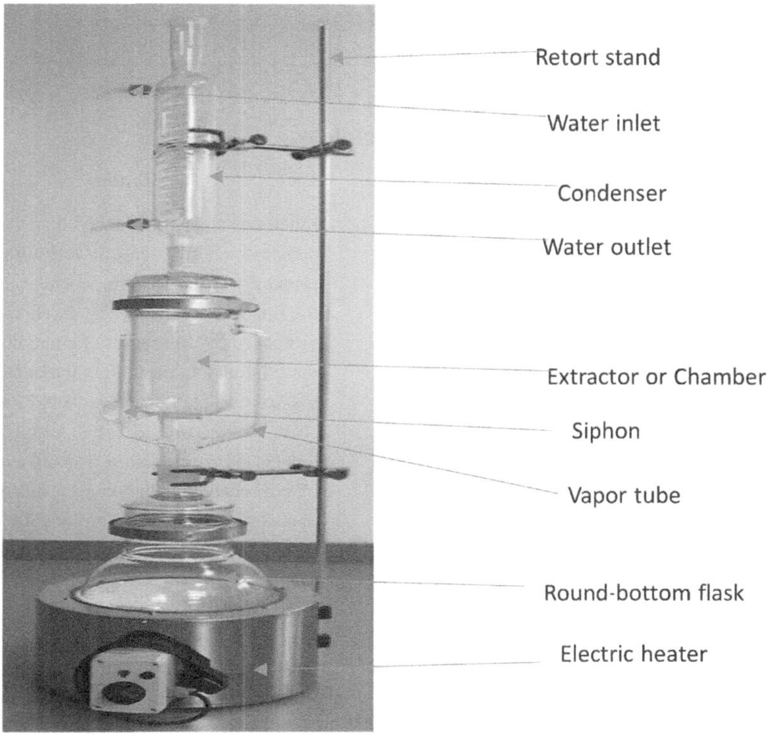

FIGURE 3.1 Soxhlet extractor laboratory apparatus setup.

Source: Photograph by the authors.

2. *Soxhlet extractor or chamber* in which the plant sample to be extracted is packed. Sometimes, the crude plant sample is packed in a sealed filter paper, placed in a cellulose thimble, and covered with cotton wool, which is placed in the Soxhlet chamber. It has a side tube that transfers solvent vapors from the boiling flask to the condenser and a siphon tube that transfers Soxhlet extract to the boiling flask.
3. *A condenser*: The Soxhlet extractor has a component called a condenser where the solvent vapors condense back into the thimble for extraction purposes.

This reflux process continues for a predetermined time, which is a function of the solid matrix nature, its surface area, desired concentration of the targeted analyte, and the solvent recovery efficiency. The desired concentrated phytoconstituents leached into the solvent is later collected by evaporating the solvent.

3.6 ADVANTAGES AND LIMITATIONS

TABLE 3.1
Advantages and limitations of SE principle.

SN	Advantages	Limitations
1.	Dynamic extraction: It outperforms static methods by consistently exposing the crude plant material to a fresh, concentrated, and active solvent condensate in each cycle, thereby improving extraction efficiency (Arceusz et al., 2013; Ciuca & Racovita, 2023).	Siphon-dependent time delay: It demonstrates an extended cycle time, primarily attributed to its dependence on the siphon mechanism. The duration is prolonged by the process of emptying the solution into the boiling flask and the gradual evaporation and condensation steps (Ciuca & Racovita, 2023)
2.	Higher and total yield: It gives higher and total recovery of yield (Kapadia et al., 2022; Saini et al., 2021).	High energy consumption: The utilization of solvents with high boiling points results in a slow reflux process, prolonging extraction times and increasing the risk of thermal degradation (Kapadia et al., 2022).
3.	Maximum extraction rate: The continuous reflux of solvent at a consistent potency contributes to high extraction efficiency, enabling a faster extraction process with reduced time requirements compared to other conventional methods (Saini et al., 2021).	Inefficiency with fully soluble plant material: It demonstrates inefficiency when handling fully soluble plant material, making a simpler filtration method more effective in such cases (Khalil et al., 2017).
4.	It serves as a precursor and reference point for newly invented extraction techniques, and its combination with certain methods, such as microwave-assisted extraction (MAE), has been proven excellent (Kapadia et al., 2022).	Hazardous solvent use: The method promotes the use of substantial amounts of hazardous and flammable organic solvents, presenting potential risks of toxic emissions during the extraction process (Saini et al., 2021).
5.	Solvent consumption reduction: It minimizes the overall solvent usage by implementing recycling. The reflux principle guarantees that evaporated solvent condenses in the Soxhlet chamber, facilitating the extraction process and subsequently returns to the boiling flask (Ciuca & Racovita, 2023).	The entire procedure is labor-intensive and time-consuming (Arceusz et al., 2013).
6.	Continuous reflux feature: Continuous reflux eliminates the necessity for frequent solvent changes. A siphon facilitates reflux, enabling condensate overflow in the Soxhlet chamber to return to the round-bottom flask without requiring additional power, except for heating (Arceusz et al., 2013).	High solvent cost: It frequently demands an appropriate and high-quality solvent for a specific extraction process (Arceusz et al., 2013; Kapadia et al., 2022).

3.7 ESSENTIAL OIL EXTRACTION PROCESS

Extraction process using Soxhlet extractor is a physicochemical means of leaching the desired essential oil that contain both volatile and nonvolatile compounds from the crude plant's parts.

3.8 SELECTION OF PLANT MATERIAL

The extraction process involves plant parts such as flowers (e.g., orange, pink, lavender), leaves (e.g., eucalyptus, mint, thyme, bay leaf, savory, sage, pine needles, and tree underground organs like vetiver), rhizomes (e.g., ginger, sweet flag), seeds (e.g., rambutan, carvi, coriander), fruits (e.g., fennel, anise, citrus epicarps), and wood and bark or peel (e.g., rambutan, cinnamon, sandalwood, rosewood) (Dhifi et al., 2016). The efficacy of isolating phytoconstituents is contingent upon meticulous selection and collection of plant materials, taking into account the plants' health status. Procedural guidelines for the collection of plant materials, encompassing pertinent collection time, techniques, and processing and storage procedures, were outlined by the US Department of Agriculture Natural Resources Conservation Service in 2011 (U.S. Department of Agriculture, Natural Resources Conservation Service (USDA-NRCS), 2011).

3.9 PRETREATMENT OF RAW MATERIALS

1. *Drying of plant materials*: The effectiveness of the extraction process and the preservation of bioactive or chemical compounds are contingent upon the drying method and conditions employed. Numerous researchers advocate for an air-drying process conducted in shaded, dark rooms to mitigate the risk of overheating, which could lead to the evaporation of volatile substances from plant materials. Additionally, exposure to sunlight during drying may result in the loss of light-sensitive constituents. Alternative drying methods encompass microwave drying, oven drying, silica gel or salt drying, and freeze-drying processes (Gopalasatheeskumar, 2018).
2. *Size reduction of plant materials*: Size reduction in the context of plant material processing involves techniques such as grinding and milling, utilizing equipment like the hammer mill, fixed head mill, plate mill, roller mill, and cutter mills. This mechanical refinement enhances the contact interactions between the ground plant particles and the extraction solvent, facilitating the extraction process (Das et al., 2010). The extraction efficiency, inherently linked to the available surface area, experiences improvement as the reduction in particle size leads to a larger surface area of the ground particles, thereby enhancing the efficiency of the extraction process (Gopalasatheeskumar, 2018).

3. *Sieving*: Sieving or size separation plays a pivotal role in enhancing the efficiency of SE, as substantiated by various researchers, as outlined in Table 3.1. Maintaining a uniform particle size post grinding is crucial, as it ensures optimal extraction by allowing the solvent to permeate uniformly through the packed ground particles in the thimble. The process of size separation involves sieving, while the determination of particle size is accomplished through microscopic analysis (Rahmalia et al., 2015).

3.10 SELECTION OF SOLVENTS

The selection of a solvent is influenced by factors including the plant type, specific plant part, bioactive compound characteristics, and solvent availability. Typically, polar solvents like water, methanol, and ethanol are employed for extracting polar compounds, while nonpolar solvents like hexane and dichloromethane are utilized for nonpolar compounds. The term "menstruum" refers to a suitable solvent applied in the extraction of medicinal plants (Abubakar & Haque, 2020). Factors affecting selection of extraction solvents include:

- Selectivity: Solvent choice should prioritize the ability to extract the desired active constituent while leaving insoluble solids.
- Safety: Optimal solvents should exhibit minimal toxicity and flammability.
- Cost-effectiveness: Solvents should be economically viable for practical extraction processes.
- Reactivity: Chosen solvents must be inert to prevent undesired reactions with the extract.
- Ease of recovery: Solvents should allow straightforward separation from the dissolved analyte through filtration or evaporation.
- Viscosity: Lower viscosity is preferred for enhanced penetration and interaction.
- Boiling temperature: Solvents with lower boiling points are desirable to prevent degradation due to heat.

3.11 DETAILED EXTRACTION PROCEDURE

The plant part for extraction samples must be fresh and healthy, which should be washed using water and soap to remove traces of soil, peeled and grinded. This is collected and stored in a refrigerator ready for the extraction.

3.11.1 LOADING THE EXTRACTOR

The prepared and refrigerated rhizome samples were allowed to thaw and then loaded on the thimble in the Soxhlet chamber. The chamber was later mounted on the round-bottom flask and the condenser was fixed to the Soxhlet chamber. This is illustrated in Figure 3.2.

TABLE 3.2
SE: particle size range, solvent type, and percentage yield.

Authors	Plant name	Plant part	Solvent type	Particle size range	Other parameters	Percentage yield/ constituents
Al-Areer et al. (2023)	Ginger	Rhizome	Water, methanol, chloroform, ethanol, ethyl acetate, *n*-hexane, dichloromethane	Coarse dried powder	8 h/constant temperature	15.9, 11.2, 5.8, 5.5, 4.5, 4, 3.2, respectively
Joshi & Cherekar (2021); Sahne et al. (2016)	Turmeric	Rhizome	Acetone	Fresh blended extract	7 h/56.53°C; 8 h/60°C	4.09/6.9
Kanadea & Bhatkhandeb (2016)	Ginger	Rhizome	Acetone, *n*-hexane, ethanol, methanol, water	Coarse powder	1 h/80°C constant	37, 16.6, 16, 15.6, 15.1, respectively
Qin & Zhong (2016)	Avocado	Fruit pulp	Hexane	Coarse powder	70°C/≤100 mmHg	54
Qin & Zhong (2016)	Avocado	Fruit pulp	Hexane	Paste	Microwave-assisted	97
Ciuca & Racovita (2023); Dutta (2015)	Curcumin	Peels	Ethanol	Coarse powder/sieve mesh 120	48 h/40°C	1.25
Tesfaye & Tefera (2017)	Neem	Seed	Ethanol-hexane 60:40%	355 μm particle size	3 h	43.71
Nebolisa et al. (2023)	*Moringa oleifera*	Seed	Petroleum ether	Coarse powder sieve mesh 0.7 mm	75°C; 4 h	28.75
Thilakarathna et al. (2023)	Mahua plant	Seed	*n*-hexane	Coarse powder	80°C/8 h	42.72
Wu et al. (2023)	Agarwood	Leaf	*n*-hexane	Coarse powder/ 24 mesh sieve	3 h	5.02
Choudhury et al. (2023)	Cumin	Seed	*n*-hexane	Coarse powder	40–60°C/8h	34.24
Singh & Kumar (2023)	Pumpkin (*Cucurbita moschata* Duch.)	Seed	*n*-hexane	Coarse powder/1 mm	40–60°C/6 h	38.03

TABLE 3.3
Comparative analysis of solvents selection for active phytoconstituents.

Solvent type	Polarity	Advantage	Disadvantage	Solubility	Phytoconstituents
Water	1.000	Wide-range usage, cheap, nontoxic, nonflammable, highly polar	Allergic to bacterial, mold growth and hydrolysis. High heat required to concentrate the extract	All polar compounds	Anthocyanins, starches, tannins, saponins, terpenoids, polypeptides, and lectins
Methanol Ethanol Butanol	0.762 0.654 0.586	Preservative ability at >20% concentration. Nontoxic at low concentration, and low heat for extract concentration	Inability to dissolve fats, gums, and wax; flammable and volatile.	Polar, miscible with water and used for secondary metabolites	Tannins, polyphenols, polyacetylenes, flavonoids, terpenoids, saterols, and alkaloids
Acetone	0.355	Colorless, less toxic, easily evaporated, flammable, and water-soluble	Highly inflammable	Polar	Tannins, polyphenols, polyacetylenes, flavonoids, terpenoids, sterols, and alkaloids
Dichloromethane Chloroform	0.309 0.259	Colorless with sweet smell, and is soluble in alcohols.	Sedative and carcinogenic	Nonpolar, soluble in alcohol, but dichloromethane is relatively easy to recover	Alkaloids, terpenoids, coumarins, fatty acids, and flavonoids
Ethyl acetate	0.228	Low boiling points, tasteless, stability	Highly volatile and flammable	Nonpolar, soluble in water	Alkaloids, terpenoids, coumarins, fatty acids, and flavonoids
Petroleum ether/ diethyl ether	0.117	35–60°C boiling point. Highly volatile, hence it is easily distilled	Less effective due to oxidation and polymerization of the oil during storage. Highly flammable	Nonpolar, insoluble in water	Alkaloids, terpenoids, coumarins, fatty acids, and flavonoids
n-hexane	0.009	Petroleum odor, 69°C boiling point and more effective in extraction	Highly volatile, flammable and reactive	More nonpolar, insoluble in water	Alkaloids, terpenoids, coumarins, fatty acids, and flavonoids

Source: Abubakar & Haque (2020).

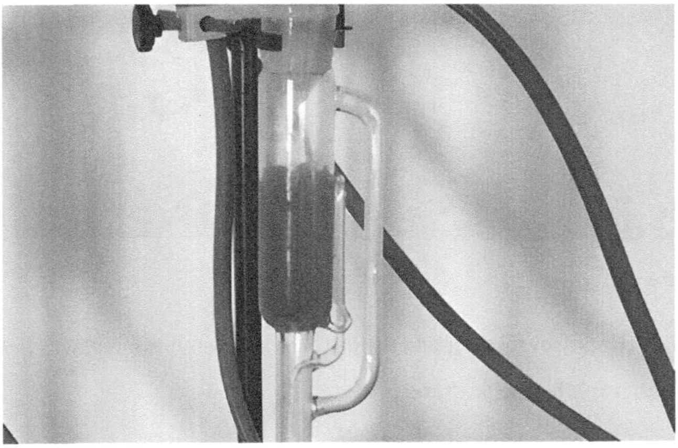

FIGURE 3.2 Soxhlet extractor/chamber loaded with solid matrix.

Source: Photograph by the authors.

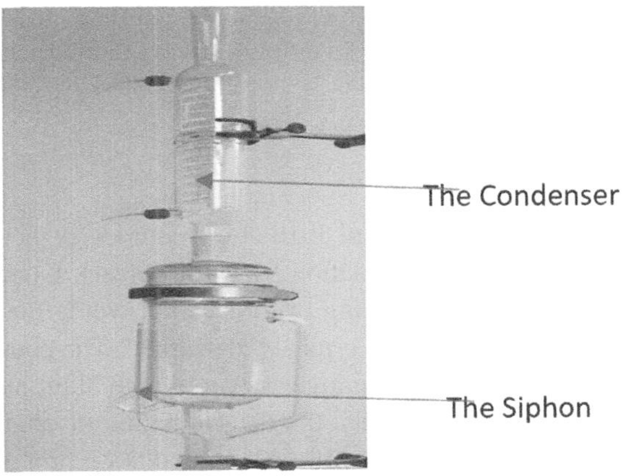

FIGURE 3.3 SE procedure showing the chamber with siphoning and the condenser.

3.11.2 Siphoning and Condensation

The solvent, maintained at a constant temperature above its boiling point in the round-bottom flask, undergoes vaporization through the side tube of the Soxhlet chamber and subsequently moves to the condenser (Figure 3.3). A water flow system within the condenser aids in the solvent vapor reflux into the Soxhlet chamber, housing the crude plant material. In this process, certain bioactive compounds dissolve in the warm solvent, inducing a gradual change in solvent color. The solution volume within the Soxhlet chamber rises, surpassing the siphon arm height, triggering the automatic runoff into the round-bottom flask.

FIGURE 3.4 Rotary evaporator and the distillation column assemblage.

Source: Photograph by the authors.

This concludes the reflux cycle, which can be iterated numerous times, spanning 6 to 12 hours or even days for exhaustive extraction. Throughout each cycle, a fraction of the nonvolatile bioactive compound dissolves in the solvent. After multiple cycles, the target bioactive compound becomes concentrated in the round-bottom flask. If the desired chemical compounds exhibit significant solubility in the solvent, rotary evaporation may be employed to separate them from the insoluble material.

3.11.3 Solvent Recovery

Commonly, the rotary evaporator and distillation methods, depicted in Figure 3.4, are employed for the extraction/isolation from the solvent. Concentration of the extract is achieved through air-drying or desiccation, while the insoluble crude plant material within the thimble is typically discarded. The concentrated extract is then stored in a tightly sealed container, covered with aluminum foil, and preserved in a refrigerator. Additionally, the determination of crude fat or oil content can be conducted using a straightforward calculation procedure outlined in Section 3.26.

3.12 FACTORS AFFECTING EXTRACTION EFFICIENCY

The following factors affect Soxhlet solvent extraction process and the extract antioxidant activity, hence they should be considered while selecting the method for extraction and for further research studies.

1. *Type of solvent (polarities) and its concentration*: Ezez and Tefera (2021) demonstrated that the ginger rhizome exhibited the highest concentration of phenolic compounds when subjected to solvents with high polarity. The extraction efficiency of phenolic content followed the order of methanol > ethanol > ethyl acetate > acetone. Furthermore, Al-Areer et al. (2023) utilized seven solvents with varying polarity to

extract essential oil from both fresh and powdered ginger. The solvents and their corresponding percentage yields for powdered ginger extract are as follows: ethanol/5.5; methanol/11.2; ethyl acetate/4.5; water/15.9; dichloromethane/3.2; chloroform/5.8; and *n*-hexane/4.0. The experimental results suggest that polar solvents tend to exhibit higher extraction efficiency compared to nonpolar solvents.

2. *Temperature*: Resham Kanadea and D. S. Bhatkhandeb (2016) investigated the effect of temperature on ginger oil yield using acetone as the solvent, revealing that higher temperatures were associated with lower ginger oil yields. In contrast, Cha et al. (2020) found that the highest extraction yield was achieved at 70°C for 90 minutes using a solvent with 35% ethanol concentration, while the lowest yield occurred at 50°C for 60 minute with a solvent having a 70% ethanol concentration. The disparity in results may be attributed to variations in solvent selection.
3. *Extraction time*: Optimal contact and residence time during extraction directly contribute to maximizing oil yield, ensuring continuous interaction between the solvent condensate and the plant parts sample within the thimble. Suwari et al. (2017) observed a 50% increase in oil yield with a 2-hour extension in extraction time using Soxhlet solvent extraction on the seed kernel of Feun Kase (*Thevetia peruviana*). Similarly, Al-Areer et al. (2023) demonstrated progressive ginger oil yield with increasing interaction time, achieving 1.27% in 1 hour, 7.52% in 1.5 hours, and 9.27% in 2 hours.
4. *Solvent-to-sample ratio*: The response surface methodology (RSM) demonstrated a significant linear effect of sample weight on oil recovery. The study concluded that the optimal solvent-to-sample ratio for oil recovery was 750 ml of methanol volume to 30 g weight of the sample, with an extraction time of 4 hours (Mohammed et al., 2022).
5. *Agitation speed*: There was a continuous rise in oil yield with an increase in agitation speed until it reached 100 rpm, after which a consistent and gradual decline in oil yield was observed. The extraction percentages were 6% at 50 rpm, 8% at 100 rpm, and 8% at 150 rpm (Panchal et al., 2014).
6. *Particle size of solid matrix*: A broader surface area of the solid matrix facilitates solvent penetration and enhances interaction, ensuring higher efficiency. This was demonstrated by Al-Areer et al. (2023) in a study using seven different solvents for the extraction of essential oil from both fresh (F) and dried powdered (P) ginger extract. Comparative results showed variations in percentages: ethanol (3.3% F, 5.5% P); methanol (2.7% F, 11.2% P); ethyl acetate (2.3% F, 4.5% P); water (1.7% F, 15.9% P); dichloromethane (0.5% F, 3.2% P); chloroform (0.4% F, 5.8% P); and *n*-hexane (0.3% F, 4.0% P).
7. *pH value*: In practical applications, it is recommended to maintain the pH of the solution of bioactive compounds lower than their pKa values to obtain the unionized forms, which exhibit a higher tendency to distribute

into the organic phase. Adjusting the pH of the solution decreases solubility in water and increases extraction efficiency. A study conducted by Orooji et al. (2021) to investigate the effect of varied pH values on the extraction efficiency recorded the highest recovery at a lower pH value of 2, with a decrease observed at higher pH values.

3.13 SAFETY CONSIDERATIONS

Soxhlet solvent extraction deals with heating of some toxic solvents and handling of glasswares, hence necessary safety measures must be adhered to when considering using the extraction method. These include;

1. Loading the sample with filter cellulose/paper filters, not metal sieve filters, is recommended to prevent the decomposition of some pesticide targets. Additionally, matching the filter cellulose/paper filter with the extractor after loading the sample is crucial to avoid blocking the siphon return tube.
2. The Soxhlet chamber/extractor used in the experiment should not be excessively large, as this may cause the solvent vapor to reach the extractor before the condensation of ambient air, reducing extraction efficiency and prolonging extraction time.
3. Experiments involving toxic solvents must always be conducted in a fume hood to prevent any potential escapes.
4. Personal protection measures, such as wearing nitrile gloves, goggles for eye protection, lab coats, and covered shoes, should be strictly adhered to.
5. Due to the relatively open nature of SE, precautions should be taken to prevent contamination during the extraction operation, including covering the top of the condenser tube during experimental procedures.
6. Cleaning the Soxhlet extractor with chromic acid washing solution or deionized water and subsequent air-drying is recommended.

3.14 TYPES OF SOXHLET APPARATUS

3.14.1 Traditional Glass Soxhlet Apparatus

The classic glass Soxhlet apparatus is a straightforward laboratory instrument invented by Franz von Soxhlet in 1879, designed for the extraction of lipids from solid materials. It consists of three compartments: a round-bottom glass flask for solvent boiling, an extractor chamber housing the thimble and connected to a siphon, and a condenser to condense vaporized solvent back into the chamber. The apparatus operates on a simple reflux and siphoning principle, ensuring continuous cyclic interaction between the solvent and the solid material in the thimble until the desired solute concentration is achieved in the boiling flask. Despite being a traditional technique, it serves as a benchmark for evaluating the effectiveness of modern extraction methods. This method offers advantages such as solvent recycling, the use of a single solvent batch, and relative cost-effectiveness. However, it is not suitable for thermolabile compounds due to the risk of compound degradation with prolonged heating hours.

3.14.2 Modern SE Systems

To mitigate the limitations of traditional SE, such as large extractant volumes and prolonged extraction times, modern SE systems have been developed over 40 decades now, which include automated Soxhlet, ultrasound-assisted extraction (UAE) and microwave-assisted SE.

3.14.2.1 Automated SE

This first version, which was also known as "Soxtec", was introduced in 1974 by Edward Randall to significantly reduce extraction time by tenfold (Majors, 2006). It was initially implemented on the commercial equipment Soxtec® System HT; other similar commercialized systems include SoxtecTM 2050, SoxtecTM 2055, SOXTHERM, and the SoxCapTM 2047. The automated Soxhlet technique (Figure 3.5) involves fully immersing the solid sample housed in the thimble or cup in the boiling solvent, increasing the extraction efficiency of analytes due to

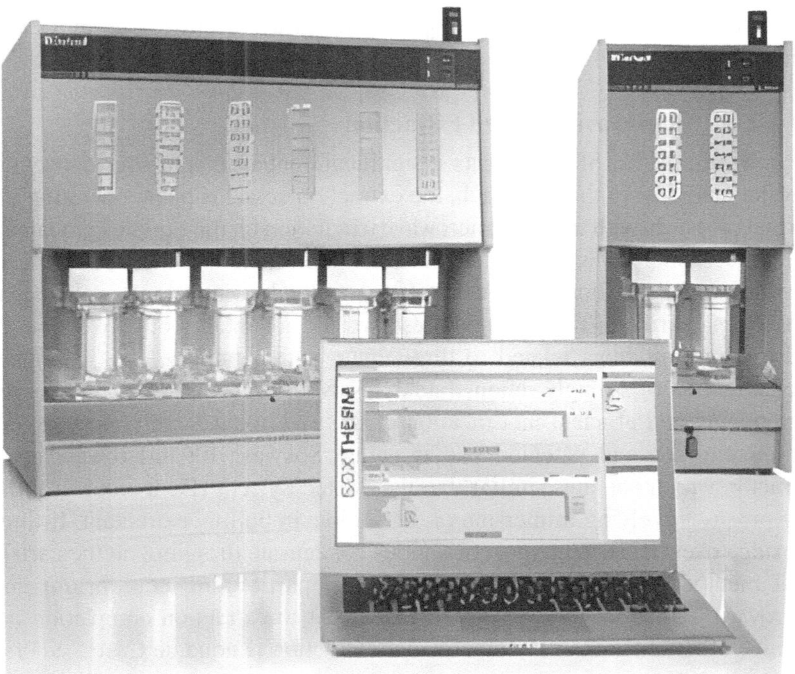

FIGURE 3.5 Automated Soxhlet extractor (SOXTHERM) commercialized by Gerhardt.

direct contact with the hot solvent. Subsequently, a rinsing period follows, flushing residual extract from the preheated solid sample with the condensed solvent inside the raised thimble. Finally, the solvent in the solution flask is evaporated to concentrate the analyte for further processing, while the preheated solid sample in the thimble or cup is allowed to air-dry. This prevents the sample from overheating, boiling dry, or potential oxidation. Simultaneously, the condensed solvent is collected, along with the evaporated solvent, in a reservoir for potential reuse or disposal. Over time, this modern system has undergone qualitative modifications, incorporating auxiliary energies and automation, leading to commercialized variants like automated SE, focused microwave-assisted SE, ultrasound-assisted SE, high-pressure SE, and fluidized bed extraction (Majors, 2006; Zygler et al., 2012; Luque de Castro & Priego-Capote, 2010).

3.14.2.2 Ultrasound-Assisted Extraction

The ultrasound-assisted Soxhlet extractor integrates ultrasound into the traditional SE process. In this setup, the Soxhlet chamber is placed in a thermostatic bath, and ultrasound is applied through an ultrasonic probe. The ultrasound induces a decompaction effect, eliminating the need for repetitive grinding between cycles. This innovative approach reduces the number of Soxhlet cycles required for efficient extraction of the targeted essential oil from the sample matrix. However, caution should be taken while setting the operating conditions for the extraction system in order to avoid the potential oxidative effects associated with drastic ultrasound conditions, which degrade the extracted oil (Luque de Castro & Priego-Capote, 2010).

3.14.2.3 Microwave-Assisted Extraction

Microwave-assisted SE represents a significant enhancement over conventional Soxhlet methods. In this approach, the extraction vessel is open, operating under normal pressure, with focused microwave irradiation on the sample compartment. The process maintains the contact between the sample and the fresh extractant, resembling the traditional Soxhlet technique. Notably, it eliminates the need for subsequent filtration. This method preserves the advantages of conventional SE, addressing limitations related to throughput, automatability, and the quantitative extraction of strongly retained analytes. Microwave-assisted SE emerges as a promising and efficient modification of the traditional Soxhlet approach. The commercialized and patented one is called Soxwave-100 microwave-assisted extractor, which operates similarly to the Soxtec System HT. It employs a three-step process involving immersion of the sample in boiling extractant, lifting the cartridge over the solvent, and continuous condensate dropping on the cartridge. This method establishes a matrix-extractant partitioning equilibrium during microwave radiation, followed by displacement to extraction completion in the absence of irradiation, ensuring efficient extraction (Luque de Castro & Priego-Capote, 2010).

3.15 COMPARISON AND SELECTION BASED ON LITERATURES AND APPLICATIONS

TABLE 3.4
Comparison and selection based on literatures and applications.

Features for selection	Traditional glass Soxhlet apparatus	Modern SE systems
Total extraction time	18 h	>2 h
Extractant solvent volumes	100 ml *n*-hexane/acetone (1:1, v/v)	50 ml *n*-hexane/acetone (1:1, v/v)
% Yield recovery	Lower	Higher
Automation	Partial	Fully
Safety and acceptability	Safe but less preferred due to the drawbacks	Safer and generally accepted, even endorsed by US Environmental Protection Agency (EPA).
Solvent selection	Selective	Most solvents may be used except diethyl ether.
Application for extraction	Limited to thermally stable analytes	Robust in application

Source: Majors (2006); Sporring et al. (2005).

3.16 APPLICATIONS OF SE IN ESSENTIAL OIL PRODUCTION

SE, originally devised by Franz von Soxhlet in 1879 for extracting fats from milk solids, has evolved over time and found applications beyond its initial scope. It has been enhanced and adapted for extracting phytoconstituents and essential oils from various solid matrices in both small- and large-scale production settings. This method is particularly useful when the target thermophilic compound has limited solubility in the chosen solvent, while impurities in the matrix remain insoluble (Zygler et al., 2012). The subsequent sections will delve into different aspects of SE, particularly its application in the production of essential oils.

3.16.1 Case Studies and Examples

Volatile components of essential oils were extracted from Pu-erh ripe tea using the SE method, along with other techniques. Pu-erh ripe tea, renowned for its woody and chestnut fragrance, is a popular Chinese tea product. To ensure sustainability and prevent waste from excess production, the tea was processed into tea essential oil. The SE involved loading 30 g of tea powder into a Soxhlet extractor thimble, using 450 ml of dichloromethane as the extractant volume, and boiling for

8 hours in three cycles. The dichloromethane was then evaporated, and the essential oil was filtered and analyzed using gas chromatography-mass spectrometry (GC-MS). The SE method yielded the highest essential oil content at 0.81 ± 0.01 g/kg, containing approximately 40 volatile compounds out of 96 identified. However, it was noted that SE had the longest extraction time (Gao et al., 2017).

In another case study, essential oil was produced from Pumpkin (*Cucurbita moschata Duch.*) seeds of 'Kashi Harit' cultivar. Various methods, including SE, were employed for oil extraction, and the results showed that SE had the highest oil yield at 38.03%. The grounded 1-mm-particle-size powder was treated with different solvents for 6 hours within the boiling range of 40–60°C, with *n*-hexane providing the best oil yield recovery (Singh & Kumar, 2023).

Simultaneous extraction of sunflower oil from its seeds and active compounds from olive leaves was conducted using SE and pressurized propane. The aim was to enhance the antioxidant activity of sunflower oil by incorporating bioactive compounds from olive leaves. SE yielded better results both individually and simultaneously, with mass percentage yields of 46.8±0.6%, 3.1±0.1%, and 50.0±0.2% for sunflower oil, olive leaf extract, and simultaneous extraction, respectively (Jaski et al., 2022).

In a study by Rajesh et al. (2023), geranium oil extraction performance was investigated using SE with various organic solvents such as *n*-hexane, methanol, chloroform, toluene, and water. Geranium (*Pelargonium graveolens*) oil is known for its antimicrobial activities. Among the solvents tested, *n*-hexane and toluene provided acceptable geranium oil yields ranging from 0.08% to 0.1%.

Last, a comparative study was conducted to assess the antioxidant, anticancer, and antimicrobial activities of essential oils from Semen Platycladi. Three extraction methods, including SE, were employed. SE, using ethanol as a solvent heated at 60°C for 6 hours, resulted in an essential oil yield of 32.45%. Although SD extraction had the lowest yield at 0.16%, the essential oils extracted through this method exhibited the best antioxidant and antimicrobial activities (Zhu et al., 2020).

3.16.2 Comparison with Other Extraction Methods

The SE technique, often considered traditional, remains the benchmark against which all other extraction procedures are evaluated. Despite its simplicity, minimal user involvement, effectiveness with a guaranteed 100% extract recovery, and cost-effectiveness, drawbacks such as prolonged extraction times and high extractant volumes exist. However, the introduction of modern SE systems has addressed the time-related limitations, enhancing the acceptance of this ancient method, even acknowledged by US EPA (Majors, 2006). Table 3.5 provides a summary of the comparison between the SE method and others.

TABLE 3.5
Summary of the comparison between the SE method and others.

Authors	Findings	Remarks
Nebolisa et al. (2023)	The impact of MAE and SE techniques on the physicochemical characteristics of *Moringa oleifera* seed oil and proteins were evaluated. The results from MAE and SE revealed oil yield percentages of 34.25 ± 0.0% and 28.75 ± 0.0%, respectively, with low moisture content measured at 0.008 ± 0.0% and 0.011 ± 0.0%. The extracted oil exhibited non-drying and unsaturated properties, being moderately saponified and less dense, with density values of 0.91 ± 0.01 g ml^{-1} and 0.92 ± 0.02 g ml^{-1}.	Both MAE and SE demonstrated feasibility and viability for the high-throughput extraction of *Moringa oleifera* oil. The selection between the two techniques can be influenced by factors such as availability and cost.
Rajesh et al. (2023)	The extraction performance of geranium oil using SD and SE methods was investigated with the goal of maximizing yield while minimizing costs and facility requirements. The results from SD and SE are as follows: - Oil yield: 0.15–0.2% for SD and 0.08–0.1% for SE. - Operating temperature: 100–115°C for SD and 65–70°C for SE. - Energy required: 42.4 kw-h for SD and 75.83 kw-h for SE. - Solvent-to-material ratio: 0.5 (v/w) for SD and 6 (v/w) for SE.	Certainly, while SD showed better performance compared to SE, the observed performance gaps may not justify the associated cost implications incurred in SD.
Thilakarathna et al. (2023)	The study on *Madhuca longifolia* seed oil extraction compared UAE, mechanical pressing (MP), and SE methods. Under mild conditions of 35 minutes, 35°C, 40% amplitude, a solvent mixture of isopropanol to acetone (1:1), and a solvent-to-sample ratio of 20 ml/g, UAE achieved approximately 56.97% oil yield and 99.54% oil recovery. The results indicate that UAE outperformed both SE and MP in terms of oil yield and recovery.	The UAE technique emerges as a promising, energy-efficient, and environmentally friendly method for extracting antioxidant-rich mahua seed oil. The study suggests that there is a need for scaling up this technique to the industrial level.

(Continued)

TABLE 3.5 (*Continued*)
Summary of the comparison between the SE method and others.

Authors	Findings	Remarks
Gao et al. (2017)	An investigative study was conducted to assess the efficiency of four different extraction methods for obtaining the volatile components of essential oils from Pu-erh ripe tea. The extraction methods included SE, UAE, simultaneous distillation extraction (SDE), and SD. The results analysis revealed the essential oil yield and number of volatile compounds for each method as follows: SE: 0.81±0.01g/kg and 40; UAE: 0.36±0.01 g/kg and 38; SDE: 0.11±0.01 g/kg and 35; and SD: 0.59±0.01 g/kg and 47.	The SD method was deemed viable and feasible for both the essential oil extraction rate and oil flavor in the conducted study.
Singh & Kumar (2023)	The investigation into the effects of cold press extraction (CPE), SE, and mechanical shaking extraction (MSE) methods on extracting phytochemicals and bioactive compounds from pumpkin seeds revealed that the SE method exhibited the highest oil yield at 38.03%, followed by CPE (33.25%) and MSE (26.36%). SE also demonstrated the highest values for acid value, peroxide value, color, and unsaponifiable matter. Conversely, the CPE method showed the highest values for all bioactive compounds, total phenolic compounds, and 2,2-diphenyl-1-picrylhydrazyl activity, while the MSE extraction displayed intermediate values for almost all properties.	The CPE method was deemed viable and feasible for oil extraction, particularly in terms of oil functional quality. However, it was noted to have lower oil throughput. The study suggests that exploring other novel hybrid methods, such as ultrasonication, could be considered for achieving total recovery in the extraction process.

3.17 ANALYSIS AND CHARACTERIZATION OF EXTRACTED ESSENTIAL OILS

Recent advancements in the study of phytoconstituents, such as essential oils and bioactive compounds, can be attributed to their unique characteristics, including antimicrobial, antioxidant, and nutraceutical properties, as well as their role in enhancing aroma and sensory quality. These distinctive features have led to a broad range of industrial applications, spanning sectors such as food, pharmaceuticals, cosmeceuticals, and fragrance and flavor industries. Consequently, it becomes crucial to uphold rigorous quality control measures and employ reliable analytical techniques for each extraction batch (Waseem & Low, 2015).

3.17.1 Quality Control Methods

Essential oils are anticipated to be natural and pure, with their quality influenced by various factors, including production technology and equipment, weather and geographical conditions, agro-biological factors, as well as the duration and storage conditions. Quality control of essential oils involves various examinations and tests to ensure their purity and adherence to standards. These include:

1. *Organoleptic assessment*: Preliminary examinations involve assessing sensory characteristics like odor, taste, color, and appearance. This organoleptic assessment compares the sample characteristics with established standards to validate purity.
2. *Physical or instrumental measurements*: This category includes various measurements such as optical rotation (interaction with polarized light), solubility, refractive index (speed of light passage in the oil), melting point (temperature transition from solid to liquid), and specific gravity (relative density). Deviations from established standards in these measurements may indicate adulteration.
3. *Chemical tests*: Laboratory techniques are employed for chemical tests to identify the chemical components in an essential oil. These tests, including aldehyde content, total alcohol, ester content and acid value, help determine the purity level by comparing the chemical components with known standards.

3.17.2 Analytical Techniques

Chromatography is an analytical technique used to separate components within a mixture by exploiting their differential migration between a stationary phase and a mobile phase. Various forces contribute to this separation, including ion exchange, partitioning, molecular sieving, and absorption. The interaction of the mixture sample with these forces determines the differential migration of individual components in the chromatographic system. The separation forces involved can be both physical, such as electrostatic interactions between ions and dipoles or Vander Waals forces, and chemical in nature. Chromatography plays a crucial role in analytical chemistry, allowing for precise separation and identification of components in complex mixtures.

1. *GC-MS*: GC-MS is an advanced analytical method widely employed for the meticulous separation and characterization of chemical components within complex mixtures, particularly essential oils. This sophisticated technique involves gas chromatography for the precise separation of components, followed by mass spectrometry to ionize and analyze the separated elements. The synergy of these two processes allows for the identification of specific components by correlating their structural information with library spectra. GC-MS plays a crucial role in determining both qualitative and quantitative purity of a mixture, ensuring accurate insights into the composition of essential oils. In a specific case study conducted by Zhu et al. (2020), focusing on Semen Platycladi essential oil, the automated injection of the sample, utilizing helium as the carrier gas, facilitated the analysis of main components under well-defined conditions. This integrated approach offers a high level of sensitivity and accuracy, making GC-MS an indispensable tool for assessing the chemical profiles of essential oils in various applications, including the pharmaceutical, cosmetic, and food industries.
2. *Thin layer chromatography (TLC)*: It serves as a qualitative technique designed to identify and determine the percentage composition of specific chemical components within a mixture sample. This method involves spotting a small amount (5–10 μl) of the sample onto a silica plate, which is then adsorbed by a suitable solvent in a jar. As the solvent travels a predetermined distance (usually 10 to 12 cm), the plates are removed, dried, and treated with a suitable reagent. This process results in colored spots corresponding to individual chemicals present in the sample. The detection and quantification of these spots are typically achieved using a scanner, providing valuable insights into the impurities and composition of the mixture. As a qualitative tool, TLC contributes to the comprehensive analysis of chemical profiles, aiding in quality control and assessment of essential oils in various industries (World Health Organization, 2011).

3.17.3 Aromatogram and Chemical Composition

The aromatogram serves as a qualitative method employed to assess the antimicrobial activity of substances, particularly essential oils, against specific microorganisms. In the context of medicinal and therapeutic applications of essential oils, aromatogram analysis becomes crucial. This analysis typically involves both *in-vitro* testing, conducted in a laboratory setting using petri dishes, and *in-vivo* testing, performed on living tissues when applicable. An example of the agar disk diffusion aromatogram method, as outlined by National Committee for Clinical

Laboratory Standards (NCCLS) (1997), was utilized by Jeldi et al. (2022) in their study on the antimicrobial and antioxidant activities of *Origanum compactum* essential oils.

The antifungal activity of these essential oils was evaluated against four Candida strains, determining the minimum inhibitory concentration (MIC) and the minimum fungicidal concentration (MFC). The results indicated significant activity against all tested strains in semi-solid medium, with MIC values ranging from 216 to 234 µg/ml and an MFC of 288 µg/ml in liquid medium. Additionally, antioxidant activity was assessed using the 2,2-diphenyl-1,1,picrylhydrazil radical method, revealing a notable antioxidant effect for both wild and cultivated *Origanum compactum* essential oils. The chemical composition analysis, conducted through gas chromatography coupled with mass spectrometry, identified 12 and 27 compounds in the essential oils of wild and cultivated plants, respectively, with carvacrol, p-cymene, and γ-terpinene being the predominant components (Jeldi et al., 2022).

3.18 YIELD ENHANCEMENT AND OPTIMIZATION

Optimization of SE involves strategic decision-making and methodological adjustments aimed at improving yield production by maximizing the contact ratio between the sample and the solvent, while minimizing input resources. This optimization process is designed to enhance the efficiency of SE within the shortest possible time.

3.18.1 Yield Improvement Strategies

The followings are the experimented strategies employed to improve SE efficiency.

Elevated temperatures: The application of elevated temperatures can have a profound impact on extraction efficiency. This is attributed to several factors, including the increased solubility of target analytes, facilitation of the breakdown of analyte–matrix interactions, and promotion of analyte diffusion to the matrix surface, thereby enhancing mass transfer of organic compounds to the solvent. The heightened temperature conditions contribute to improved solvation power of solvents and result in increased extraction rates, as highlighted by de la Guardia and Armenta (2011).

Extraction in several stages: Conducting extractions in multiple stages, as illustrated by the findings of Aspé and Fernández (2011), reveals a substantial increase in extraction yield with an augmentation in the number of extraction stages, particularly up to stage 3. The relationship between yield and the number of extraction stages is characterized by a linear function, indicating a direct correlation between these variables.

Solvent combination: The exploration of solvent combinations, along with variations in extraction time and temperature, was undertaken by Suwari et al. (2017) to attain optimal yield. The experiment resulted in a maximum oil yield of 51.88±3.18% under the optimized conditions of 4.5 hours, 65°C, and a methanol-to-petroleum ether ratio of 10:90 (0.6 polarity index). This investigation highlights the significance of carefully chosen solvent combinations and operating conditions in enhancing extraction efficiency.

Extraction methods combination: Engaging more than one extraction methods also help in identifying the most effective methods to extract essential oil (Aspé & Fernández, 2011).

Particle size: The impact of particle size on extraction efficiency was investigated by Al-Areer et al. (2023), who utilized fresh and dried ginger rhizomes for essential oil extraction. The study revealed that grinding the sample into a powdered form significantly enhanced the extraction of active components compared to using fresh ginger. This underscores the importance of particle size in influencing the surface area, thereby facilitating improved penetration and interaction between the solvent and the sample during the extraction process.

3.18.2 Optimization Techniques

Optimization techniques serve as powerful tools for efficiently utilizing available resources to achieve desired design parameters under optimal operating conditions. By guiding experimental work, these techniques minimize process risks, costs, and environmental impacts. The overarching goal of experimentation is to either maximize profit or minimize costs. Optimization aids in determining values for decision variables that align with desired objectives while satisfying existing constraints. The reliability of optimal solutions is contingent upon well-formulated objective functions and the chosen optimization technique. These techniques necessitate mathematical models that describe and predict process behavior, making them invaluable for estimating unknown parameters in complex nonlinear processes, identifying uncertainty variables in dynamic processes, and facilitating the scale-up methodology and design of multiphase reactors and flow systems (Alwan, 2016). Optimization techniques are categorized into three main types based on their applications:

1. *Deterministic techniques*: These methods are employed for convex or differential equations, as well as linear and nonlinear programming involving single and multi-variable functions. Examples include quadratic programming, simplex and non-simplex algorithms, Newton and quasi-Newton methods, Secant methods, and Lagrange multipliers.
2. *Statistical techniques*: These techniques find application in correlation and experimental design. Examples include least squares, RSM, and analysis of variance.
3. *Stochastic techniques*: Applied to handle complex non-convex (non-differential) equations, stochastic techniques include genetic algorithms, pattern search algorithms, and the firefly algorithm. These methods are particularly useful when dealing with uncertainty and variability in optimization problems.

3.19 EXPERIMENTAL DESIGN IN SE

A statistical optimization technique utilized by Mohammed et al. (2022) in the extraction of essential oil from *Zingiber officinale* would be adopted in this subsection. The technique was used to design an experiment to optimize the process parameters of SE.

3.19.1 Sample Preparation, Essential Oil Production, and Analysis

The ginger rhizome underwent a series of processing steps for SE. Initially, the rhizome was sourced locally, meticulously washed, and subsequently dried in an oven at 30°C for 5 days. Following this, the dried ginger was ground into a powder, from which 10 g portions were loaded into the thimble of the Soxhlet extractor. The extraction process involved heating 250 ml of methanol as the extractant volume for 3 hours. This procedure was iteratively performed for 20 g and 30 g of the powdered sample, utilizing 500 ml and 750 ml of methanol for 4 hours and 5 hours, respectively. The selection of sample quantities adhered to a statistical design specified in the Box–Behnken standard design. The refluxing temperature was maintained at 65°C during the extraction. Following the extraction process, methanol recovery was accomplished using a rotary evaporator. Essential oil concentration, expressed as a percentage area, was determined using GC-MS.

3.19.2 Statistical Analysis Using RSM

The study employed RSM to analyze two responses, extracted oil weight percentage (Y_1) and concentration percentage (Y_2), in relation to three independent variables: extraction time (X_1), methanol volume (X_2), and sample weight (X_3) in the SE process. A total of 15 experimental runs were conducted based on the designed experiments to identify the optimal conditions for achieving maximum extracted oil weight. The second-order model revealed the interplay between the dependent and independent variables. The analysis highlighted that sample weight had a significant linear effect on oil recovery, while extraction time exhibited a major quadratic effect on the percentage area concentration of essential oil. The RSM predicted the maximum essential oil recovery and percentage area concentration at 15.2% and 22.64%, respectively. The optimum operational conditions for achieving these outcomes were determined as 4 hours extraction time, 750 ml methanol volume, and 30 g sample weight for essential oil recovery, and 5 hours extraction time, 500 ml methanol volume, and 30 g sample weight for percentage area concentration of essential oil.

3.20 ENVIRONMENTAL AND SUSTAINABILITY CONSIDERATIONS

The generation of waste in the food supply chain remains a significant concern, with approximately 39% of food production reported as being wasted during the manufacturing stage (Cristóbal et al., 2018). Addressing this issue necessitates the implementation of environmental sustainability measures aimed at valorizing these waste streams. One approach involves repurposing food waste for the production of animal feed or utilizing it in bioenergy generation through anaerobic digestion processes (Cristóbal et al., 2018). Implementing such strategies not only mitigates the environmental impact of food waste but also contributes to the development of more sustainable practices within the food industry.

3.20.1 Solvent Selection and Waste Management

The choice of extractant volume in SE is crucial, as it plays a supportive role by dissolving the solute within the solid matrix. This volume is typically recovered at the end of each extraction process, discouraging the use of toxic or environmentally harmful solvents when sustainable alternatives exist (Byrne et al., 2016). Solvent selection is influenced by factors such as polarity and volatility. Highly polar solvents can dissolve a wide range of analytes, enhancing reactions, but may pose reproductive toxicity risks (Ashcroft et al., 2015; Buhler & Reed, 1990). Volatility allows for solvent recovery through distillation but may lead to air emissions and worker exposure risks. Therefore, the quest for an effective and eco-friendly solvent for SE is paramount.

The ideology behind solvent selection aims to minimize waste in two key ways. First, selecting effective solvents enhances conversion rates and recovery efficiency, ultimately reducing waste. Second, solvent recycling and recovery efforts contribute to minimizing waste generated during the SE process. Despite the demonstrated influence of solvent selection on SE efficiency by numerous researchers, a comprehensive guide or flowchart for optimum choices is yet to be fully developed. The environmental impact and waste management aspects of solvents need careful consideration, encompassing chemical efficiency, industrial constraints, cost, environmental, health, and safety quality, before their incorporation into the SE process.

3.20.2 Green SE Practices

In an effort to phase out environmentally harmful solvents, aligning with Sustainable Development Goal 12 on Responsible Consumption and Production, a focus is placed on managing chemical wastes throughout their life cycle, adhering to international agreements to minimize their release into the air, water, and soil, thereby reducing adverse effects on human health and the environment. This initiative underscores a transition toward utilizing green solvents derived from renewable sources, marking a notable trend in sustainable practices. Green extraction methods, propelled by modern technology (Figure 3.6), are emerging, minimizing or eliminating the use of organic solvents to reduce environmental and health impacts. Hu's study exemplifies this approach, utilizing MAE to enhance the efficiency and quality of *Sapindus mukorossi* seed oil extraction. The

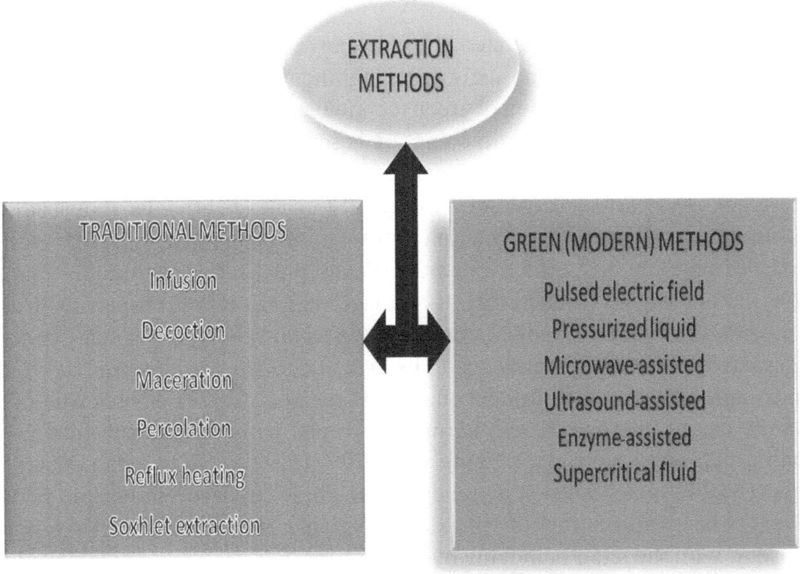

FIGURE 3.6 Modern and conventional extraction methods.

Source: Adapted and modified by the author (Panja, 2018).

optimal MAE conditions, determined through Box–Behnken design, included a mixed solvent of *n*-hexane and ethanol (4:1, v/v), microwave power of 460 W, solvent-to-material ratio of 8 ml/g, extraction temperature of 72°C, and extraction time of 42 min. The achieved oil yield of 40.12% rivaled that of SE at 40.63%. Both oils exhibited similar fatty acid profiles, triglyceride profiles, and thermal behavior. However, MAE-produced oil demonstrated superior quality, particularly in terms of low acid value and peroxide value. Notably, MAE exhibited lower solvent-to-material ratio, extraction time, energy consumption, and carbon dioxide emission compared to SE. The results concluded that MAE is an efficient and eco-friendly method for extracting oil from *Sapindus mukorossi* seeds (Hu et al., 2021).

3.21 CHALLENGES AND FUTURE TRENDS OF SE

Although considered a more traditional technique, SE yield recover efficiency remains the accepted standard for comparison with modern extraction effectiveness. However, it has been faced with the challenges of sustainability and environmental assessment. These challenges and current emerges shall be discussed in this subsection.

3.21.1 Current Challenges in SE

The current challenges associated with SE, particularly in the extraction of thermolabile compounds like capsaicin, pose significant considerations for the industrial standard and the adoption of green technology. SE, while offering high yields, faces limitations in dealing with compounds sensitive to prolonged heating, such as capsaicin, which can undergo degradation at elevated temperatures, extreme pH values, and extended extraction times (Si et al., 2014). The industrial production of capsaicin involves multiple purification stages to achieve the required purity, demanding cost-effective extraction and purification techniques that preserve the compound's nutritional, functional, and biological properties (Castro-Muñoz et al., 2022). Hexane, a commonly used solvent in capsaicin extraction, raises environmental concerns due to its lack of eco-friendliness, and its residues can compromise the final product quality (Wang et al., 2023). Recent studies suggest the use of alternative, less harmful solvents, with organic solvents demonstrating capsaicin recovery ranging from 70% to 92%. SE stands out with the highest yield among various methods, although an inverse relationship between solvent polarity and extraction time exists. Notably, SE outperforms UAE in yield recovery, with the latter achieving over 95% capsaicin recovery at increased temperatures (30°C to 45°C) (Castro-Muñoz et al., 2022). However, SE's drawbacks, including large extractant volumes and prolonged extraction times, contribute to increased overall extraction costs. The overuse of harmful solvents further complicates compliance with environmental and health guidelines. Consequently, there is a pressing need to develop more cost-effective processes and transition toward sustainable feedstocks in the quest for environmentally friendly and economically viable extraction methods (Wang et al., 2023).

3.21.2 Emerging Technologies and Innovations

SE, a staple in analytical laboratories since its inception in 1879, has undergone continuous modifications to address its inherent limitations. A notable advancement in this methodology is the Soxtec system, introduced by Edward Randall in 1974. Departing from the conventional Soxhlet approach, Soxtec achieves enhanced efficiency by immersing solid samples completely in boiling solvent within a thimble. This direct and efficient contact between the sample and the hot solvent is a key feature. Subsequent steps include a rinsing period to flush residual extracts from the preheated solid sample using condensed solvent, with the thimble raised above the boiling solvent. The solvent in the solution flask is then evaporated to concentrate the analyte, while the preheated solid sample undergoes controlled air-drying to prevent overheating, oxidation, and environmental pollution. The condensed solvent, collected with the evaporated solvent, can be reused or properly disposed of (Sporring et al., 2005; Zygler et al., 2012).

In addition to Soxtec, emerging technologies such as microwave-assisted SE, ultrasound-assisted SE, high-pressure SE, and fluidized bed extraction have gained prominence. A comparative study by Sporring et al. (2005) evaluating these modern techniques alongside traditional SE showcased their efficacy in extracting polychlorinated biphenyls from soil. These methods not only performed well but also demonstrated advantages such as reduced extraction time, extractant volume, and improved yield quality, all the while maintaining mild operating conditions. Importantly, these modern technologies align with green principles, contributing to eco-friendly processes and by-products.

3.22 FUTURE PROSPECTS

The future outlook for SE appears promising with the integration of emerging technologies that offer cost-effective, high-efficiency recovery, and improved purification of analytes from their natural matrices. The potential synergy between multiple extraction techniques, whether traditional, emerging, or novel, holds promise for enhancing the overall extraction and purification processes. Wang et al. (2021) demonstrated the effectiveness of a hybrid approach combining different extraction methods to achieve optimal recovery efficiency and purification.

Standardized guidelines for optimized extraction and purification processes, including comprehensive waste management and recycling strategies, are essential for advancing SE. Membrane-based technologies have been identified as particularly effective in achieving higher recovery and concentration of bioactive compounds and their by-products from liquid systems (Castro-Muñoz et al., 2022).

The adoption of green solvent technology presents an avenue for addressing the environmental impact of SE. Deep eutectic solvents (DESs) emerge as a promising substitute for traditional organic solvents. DESs offer advantages such as cost-effectiveness, ease of separation, versatility, lower toxicity, biodegradability, high extraction efficiency, chemical stability, non-flammability, good viscosity, and overall eco-friendliness. Their application spans various domains, including the extraction of bioactive compounds, dissolution of biocomposites, analytical determinations, and extraction of heavy metals, showcasing their versatility and potential in advancing extraction methodologies. The integration of these advancements is poised to further refine and elevate SE in terms of efficiency, environmental impact, and versatility.

3.23 SUMMARY OF KEY POINTS

1. Ensure you apply silica grease to lubricate the cork entrance of the assemblage, the contacts where the Soxhlet chamber is fixed into the condenser and the round-bottom flask to avoid cracks or difficulty when disengaging.
2. Generally, hexane is the most commonly used solvent for the extraction of oils from plant sources due to its low boiling temperature and easy recovery from the extract, and since most oils are soluble in hexane.

3. Boiling stones can also be added to the solvent to aid quick attainment of the solvent boiling point.
4. Defatting means removal of steroidal compounds, oil and fats from crude plant extract. Hexane and petroleum ether are often used to accomplish this. Defatting is essential because the waxy substances produce the emulsification process with the solvent and interferes in the extraction process.
5. The very fine and very coarse powders are not suitable for effective extraction because the very fine powder may produce the beds during extraction and very coarse powders delay the extraction process.
6. A cogent advantage of Soxhlet solvent extraction is that instead of many portions of warm solvent being passed through the sample, just one batch of solvent is recycled.
7. Solvent selection minimizes waste through increased conversion rate and recovery efficiency.
8. Environmental impact and waste management considerations include chemical efficiency, cost, health, and safety.
9. Waste valorization is crucial in the food supply chain, with approximately 39% of food production wasted during manufacturing.
10. Solvent selection in SE, considering factors like polarity and volatility, plays a pivotal role in minimizing environmental impact and waste generation.
11. Environmental sustainability measures should discourage the use of toxic or environmentally damaging solvents, emphasizing the need for effective and eco-friendly alternatives in SE.
12. Solvent selection influences SE efficiency, and there is a need for a comprehensive guide for optimal choices, considering environmental impact, cost, health, and safety.
13. Emerging technologies in analyte extraction enhance the future prospects of SE by offering cost-effective, high recovery efficiency, and purification from natural matrices.
14. Membrane-based technologies show higher recovery and concentration of bioactive compounds and by-products from liquid systems.
15. Green solvent technology, particularly emerging DESs, offers a substitute for harmful organic solvents in SE, maximizing their potential for various applications.
16. Conventional SE remains effective for solid–liquid extraction, maintaining its standard status for recovery efficiency over more than a century.
17. Soxtec system and modern technologies address drawbacks like long extraction time, extractant volume, and environmental impact.
18. Promising future applications for SE include the use of green solvents, and documented process optimization for various bioactive compounds extraction and purifications serves as a reference for future applications in laboratories and industries.

3.24 IMPORTANCE OF SE IN ESSENTIAL OIL PRODUCTION

1. It is used to extract essential bioactive compounds from various natural sources.
2. It is also used to separate oils from any solid mixture.
3. It made total extraction of essential oils possible to the tune of 100% recovery efficiency.
4. It serves as comparative standard for emerging extraction technologies.

3.25 DETAILED EXPERIMENTAL PROTOCOLS

1. Fill the round-bottom flask with the selected solvent.
2. Place the thimble containing sample matrix into the extraction tube.
3. Attach the extraction tube with flask containing solvent.
4. Attach a condenser unit with the extraction tube and run the water.
5. Assemble the Soxhlet apparatus containing the solvent on hot plate for the heating process.
6. The boiling solvent evaporates and condenses in the extraction tube.
7. Continue this process till all the fat is extracted. Then, stop the process and take out the thimble.
8. Repeat step (3) along with fat and condenser unit.
9. Concentrate the solution by continuous heating to evaporate and recover the solvent.
10. Terminate the heating process and clean the extraction tube and thimble.

3.26 SAFETY GUIDELINES

- The method is mostly used to analyze potentially hazardous samples and uses potentially hazardous reagents. Thus, all manipulations and experiments with such samples and reagents should be done under fume hood.
- Use of a respirator is advisable whenever a hood is not adequate or feasible. Splash guard goggles should be used when acid rinsing, solvent rinsing, transferring samples or when working with solvents under pressure.
- Prior to performing this analysis, review of all standards procedures for reagents and materials should be well noted. Observe the recommended safety precautions. Protective clothing and safety glasses should be worn when handling samples or reagents.
- Maintain a clean and uncluttered workspace. Return all chemicals, reagents, and resultant wastes to their designated storage area at the completion of the test.
- Calculation by the author: Calculation involving percentage determination of crude fat/oil content:

$$\%\ \text{Crude fat or oil} = \frac{W_2 - W_1}{W_S} \times 100\%,$$

where W_2 = Weight of flask + Fat content
W_1 = Weight of empty flask
W_s = Weight of sample.

3.27 CONCLUSION

For over a century, conventional SE has maintained its prominence as a standard method for solid–liquid extraction, demonstrating consistent and effective performance. The technique's robustness in achieving high percentage yield recovery through the principles of boiling and refluxing has established it as a benchmark, even in comparison with emerging technologies. The experiential studies reviewed underscore the significance of solvent selection, proper drying of solid samples, and size reduction and separation of sample materials as crucial factors influencing the efficiency of the extraction process.

Recognizing the limitations of conventional SE, such as long extraction times, large extractant volumes, and environmental impact, the advent of the Soxtec system and modern technologies has addressed these drawbacks. Soxtec, introduced in 1974, represents a notable advancement by immersing solid samples directly in boiling solvent, reducing extraction time and solvent volume. Additionally, emerging technologies, including MAE, UAE, high-pressure SE, and fluidized bed extraction, offer greener alternatives with reduced extraction times, solvent volumes, and milder operating conditions.

The future prospects for SE appear promising, especially with the incorporation of green solvent technologies. DESs stand out as eco-friendly substitutes for traditional organic solvents, demonstrating advantages such as cost-effectiveness, ease of separation, versatility, lower toxicity, and biodegradability. The integration of optimized extraction and purification processes, along with the exploration of hybrid technologies, further enhances the potential applications of SE in both laboratory and industrial settings. The documentation of these processes provides valuable references for future research and applications in phytochemical studies and beyond.

REFERENCES

Abubakar, A., & Haque, M. (2020). Preparation of medicinal plants: Basic extraction and fractionation procedures for experimental purposes. *Journal of Pharmacy and Bioallied Sciences*, *12*(1), 1. https://doi.org/10.4103/jpbs.JPBS_175_19

Aier, I., & Panwar, N. L. (2022). Development of biomass-based hydrothermal liquefaction system for essential oil extraction. *International Journal of Ambient Energy*, *43*(1), 952–957. https://doi.org/10.1080/01430750.2019.1682039

Al-Areer, N. W., Al Azzam, K. M., Al Omari, R. H., Al-Deeb, I., Bekbayeva, L., & Negim, E.-S. (2023). Quantitative analysis of total phenolic and flavonoid compounds in different extracts from ginger plant (*Zingiber officinale*) and evaluation of their anticancer effect against colorectal cancer cell lines. *Pharmacia*, *70*(4), 905–919. https://doi.org/10.3897/pharmacia.70.e103936

Alwan, G. M. (2016, October). *Optimization Technique* [Technical report]. ResearchGate.

Arceusz, A., Wesołowski, M., & Konieczyński, P. (2013). Methods for extraction and determination of phenolic acids in medicinal plants: A review. *Natural Product Communications*, *8*(12), 1821–1829. https://doi.org/10.1177/1934578X1300801238

Ashcroft, C. P., Dunn, P. J., Hayler, J. D., & Wells, A. S. (2015). Survey of solvent usage in papers published in *Organic Process Research & Development* 1997–2012. *Organic Process Research & Development*, *19*(7), 740–747.

Aspé, E., & Fernández, K. (2011). The effect of different extraction techniques on extraction yield, total phenolic, and anti-radical capacity of extracts from Pinus radiata Bark. *Industrial Crops and Products*, *34*(1), 838–844. https://doi.org/10.1016/j.indcrop.2011.02.002

Buhler, D. R., & Reed, D. J. (Eds.). (1990). *Ethel Browning's toxicity and metabolism of industrial solvents* (2nd ed., Vol. 2: *Nitrogen and phosphorus solvents*). Elsevier.

Byrne, F. P., Jin, S., Paggiola, G., Petchey, T. H. M., Clark, J. H., Farmer, T. J., Hunt, A. J., McElroy, C. R., & Sherwood, J. (2016). Tools and techniques for solvent selection: Green solvent selection guides. *Sustainable Chemical Processes*, *4*, 7.

Castro-Muñoz, R., Gontarek-Castro, E., & Jafari, S. M. (2022). Up-to-date strategies and future trends towards the extraction and purification of Capsaicin: A comprehensive review. *Trends in Food Science & Technology*, *123*, 161–171. https://doi.org/10.1016/j.tifs.2022.03.014

Cha, J., Kim, C.-T., & Cho, Y.-J. (2020). Optimizing extraction conditions for functional compounds from ginger (*Zingiber officinale* Roscoe) using response surface methodology. *Food Science and Biotechnology*, *29*(3), 379–385.

Choudhury, Md. S., Islam, Md. N., Khan, Md. M., Ahiduzzaman, Md., Masum, Md. M. I., & Ali, Md. A. (2023). Effect of extraction methods on physical and chemical properties and shelf life of black cumin (Nigella sativa L.) oil. *Journal of Agriculture and Food Research*, *14*, 100836. https://doi.org/10.1016/j.jafr.2023.100836

Ciuca, M. D., & Racovita, R. C. (2023). Curcumin: Overview of extraction methods, health benefits, and encapsulation and delivery using microemulsions and nanoemulsions. *International Journal of Molecular Sciences*, *24*(10), 8874. https://doi.org/10.3390/ijms24108874

Cristóbal, J., Castellani, V., Manfredi, S., & Sala, S. (2018). Prioritizing and optimizing sustainable measures for food waste prevention and management. *Waste Management*, *72*, 3–16.

Das, K., Tiwari, R. K. S., & Shrivastava, D. K. (2010). Techniques for evaluation of medicinal plant products as antimicrobial agents: Current methods and future trends. *Journal of Medicinal Plants Research*, *4*(2), 104–111.

Dhifi, W., Bellili, S., Jazi, S., Bahloul, N., & Mnif, W. (2016). Essential oils' chemical characterization and investigation of some biological activities: A critical review. *Medicines*, *3*(4), 25. https://doi.org/10.3390/medicines3040025

Dutta, B. (2015). Study of secondary metabolite constituents and curcumin contents of six different species of genus Curcuma. *Journal of Medicinal Plants Studies*, *3*, 116–119.

Ezez, D., & Tefera, M. (2021). Effects of solvents on total phenolic content and antioxidant activity of ginger extracts. *Journal of Chemistry*, *2021*, 6635199.

Gao, X., Lv, S., Wu, Y., Li, J., Zhang, W., Meng, W., Wang, C., & Meng, Q. (2017). Volatile components of essential oils extracted from Pu-erh ripe tea by different extraction methods. *International Journal of Food Properties*, *20*(sup1), S240–S253. https://doi.org/10.1080/10942912.2017.1295256

Gopalasatheeskumar, K. (2018). Significant role of Soxhlet extraction process in phytochemical research. *Mintage Journal of Pharmaceutical & Medical Sciences*, *7*(Suppl 1), 43–47.

de la Guardia, M., & Armenta, S. (2011). *Green analytical chemistry: Theory and practice* (*Comprehensive analytical chemistry*, Vol. 57). Elsevier.

Hu, B., Xi, X., Li, H., Qin, Y., Li, C., Zhang, Z., Liu, Y., Zhang, Q., Liu, A., Liu, S., & Luo, Q. (2021). A comparison of extraction yield, quality and thermal properties from Sapindus mukorossi seed oil between microwave assisted extraction and Soxhlet extraction. *Industrial Crops and Products*, *161*, 113185. https://doi.org/10.1016/j.indcrop.2020.113185

Hunter, M. (2009). *Essential oils: Art, agriculture, science, industry and entrepreneurship (A focus on the Asia-Pacific region)*. Nova Science Publishers.

Jaski, J. M., Abrantes, K. K. B., Zanqui, A. B., Stevanato, N., da Silva, C., Barão, C. E., Bonfim-Rocha, L., & Cardozo-Filho, L. (2022). Simultaneous extraction of sunflower oil and active compounds from olive leaves using pressurized propane. *Current Research in Food Science*, *5*, 531–544.

Jeldi, L., Ouled Taarabt, K., Mazri, M. A., Ouahmane, L., & Alfeddy, M. N. (2022). Chemical composition, antifungal and antioxidant activities of wild and cultivated *Origanum compactum* essential oils from the municipality of Chaoun, Morocco. *South African Journal of Botany*, *147*, 852–858.

Jensen, W. B. (2007). The origin of the Soxhlet extractor. *Journal of Chemical Education*, *84*(12), 1913–1914. https://doi.org/10.1021/ed084p1913

Joshi, N. D. A. A. K., & Cherekar, M. N. (2021). Extraction and purification of curcumin from turmeric. *Asian Journal of Plant Science Research*, *11*, 256–259.

Jugreet, B. S., & Mahomoodally, M. F. (2020). Essential oils from nine exotic and endemic medicinal plants from Mauritius show *in vitro* antibacterial and antibiotic-potentiating activities. *South African Journal of Botany*, *132*, 355–362. https://doi.org/10.1016/j.sajb.2020.05.001

Kanadea, R., & Bhatkhandeb, D. S. (2016). Extraction of ginger oil using different methods and effect of solvents, time, temperature to maximize yield. *International Journal of Advances in Science, Engineering and Technology (IJASEAT)*, *3*(4), 240–243.

Kapadia, P., Newell, A. S., Cunningham, J., Roberts, M. R., & Hardy, J. G. (2022). Extraction of high-value chemicals from plants for technical and medical applications. *International Journal of Molecular Sciences*, *23*(18), 10334. https://doi.org/10.3390/ijms231810334

Khalil, A. A., Rahman, U. U., Khan, M. R., Sahar, A., Mehmood, T., & Khan, M. (2017). Essential oil eugenol: Sources, extraction techniques and nutraceutical perspectives. *RSC Advances*, *7*, 32669–32681. https://doi.org/10.1039/C7RA04803C

Luque de Castro, M. D., & Priego-Capote, F. (2010). Soxhlet extraction: Past and present panacea. *Journal of Chromatography A*, *1217*(16), 2383–2389. https://doi.org/10.1016/j.chroma.2009.11.027

Majors, R. E. (2006, July 1). Modern technique for the extraction of solid materials. *LCGC North America*, *24*(7). https://www.chromatographyonline.com/view/modern-technique-extraction-solid-materials-update

Mohammed, H. H., Laftah, W. A., Noel Ibrahim, A., & Che Yunus, M. A. (2022). Extraction of essential oil from *Zingiber officinale* and statistical optimization of process parameters. *RSC Advances*, *12*(8), 4843–4851. https://doi.org/10.1039/D1RA06711G

National Committee for Clinical Laboratory Standards (NCCLS). (1997). *Performance standards for antimicrobial disk susceptibility tests* (Approved standard M2-A6). Wayne, PA: NCCLS.

Nebolisa, N. M., Umeyor, C. E., Ekpunobi, U. E., Umeyor, I. C., & Okoye, F. B. (2023). Profiling the effects of microwave-assisted and soxhlet extraction techniques on the physicochemical attributes of *Moringa oleifera* seed oil and proteins. *Oil Crop Science*, *8*(1), 16–26. https://doi.org/10.1016/j.ocsci.2023.02.003

Nerio, L. S., Olivero-Verbel, J., & Stashenko, E. (2010). Repellent activity of essential oils: A review. *Bioresource Technology*, *101*(1), 372–378. https://doi.org/10.1016/j.biortech.2009.07.048

Orooji, N., Takdastan, A., Yengejeh, R. J., Jorfi, S., & Davami, A. H. (2021). A quick and inexpensive method to determine 2,4-dichlorophenoxyacetic acid residues in water samples by HPLC. *Desalination and Water Treatment*, *217*, 329–338.

Panchal, B., Deshmukh, S., & Sharma, M. (2014). Optimization of oil extraction and characterization from *Tamarindus indica* Linn seed oil. *International Journal of Oil, Gas and Coal Engineering*, *2*(1), 1–6.

Panja, P. (2018). Green extraction methods of food polyphenols from vegetable materials. *Current Opinion in Food Science*, *23*, 173–182.

Qin, X., & Zhong, J. (2016). A review of extraction techniques for avocado oil. *Journal of Oleo Science*, *65*(11), 881–888. https://doi.org/10.5650/jos.ess16063

Rahmalia, W., Fabre, J.-F., & Mouloungui, Z. (2015). Effects of cyclohexane/acetone ratio on bixin extraction yield by accelerated solvent extraction method. *Procedia Chemistry*, *14*, 455–464.

Rajesh, Y., Khan, N. M., Raziq Shaikh, A., Mane, V. S., Daware, G., & Dabhade, G. (2023). Investigation of geranium oil extraction performance by using soxhlet extraction. *Materials Today: Proceedings*, *72*, 2610–2617. https://doi.org/10.1016/j.matpr.2022.07.276

Sahne, F., Najafpour, G. D., & Moghadamnia, A. A. (2016). Extraction of bioactive compound curcumin from turmeric (Curcuma longa L.) via different routes: A comparative study. *Pakistan Journal of Biotechnology*, *13*, 173–180.

Saini, R. K., Prasad, P., Shang, X., & Keum, Y.-S. (2021). Advances in lipid extraction methods – a review. *International Journal of Molecular Sciences*, *22*(24), 13643. https://doi.org/10.3390/ijms222413643

Si, W., Man, S. W., Chen, Z.-Y., & Chung, H. Y. (2014). Stability of capsaicinoid content at raised temperatures. *Natural Product Communications*, *9*(7), 1934578X1400900. https://doi.org/10.1177/1934578X1400900726

Singh, A., & Kumar, V. (2023). Phyto-chemical and bioactive compounds of pumpkin seed oil as affected by different extraction methods. *Food Chemistry Advances*, *2*, 100211. https://doi.org/10.1016/j.focha.2023.100211

Sporring, S., Bøwadt, S., Svensmark, B., & Björklund, E. (2005). Comprehensive comparison of classic Soxhlet extraction with Soxtec extraction, ultrasonication extraction, supercritical fluid extraction, microwave assisted extraction and accelerated solvent extraction for the determination of polychlorinated biphenyls in soil. *Journal of Chromatography A*, *1090*(1–2), 1–9. https://doi.org/10.1016/j.chroma.2005.07.008

Suwari, S., Kotta, H. Z., & Buang, Y. (2017). Optimization of Soxhlet extraction and physicochemical analysis of crop oil from seed kernel of Feun Kase (*Thevetia peruviana*). *AIP Conference Proceedings*, *1911*(1), 020005.

Tesfaye, B., & Tefera, T. (2017). Extraction of essential oil from neem seed by using Soxhlet extraction methods. *International Journal of Advanced Engineering, Management and Science*, *3*(6), 646–650. https://doi.org/10.24001/ijaems.3.6.5

Thilakarathna, R. C. N., Siow, L. F., Tang, T.-K., Chan, E.-S., & Lee, Y.-Y. (2023). Physicochemical and antioxidative properties of ultrasound-assisted extraction of mahua (Madhuca longifolia) seed oil in comparison with conventional Soxhlet and mechanical extractions. *Ultrasonics Sonochemistry*, *92*, 106280. https://doi.org/10.1016/j.ultsonch.2022.106280

U.S. Department of Agriculture, Natural Resources Conservation Service (USDA-NRCS). (2011, August). *Plant materials collection guide* (Technical Note, Plant Materials No. 8). Pacific Islands Area.

Wang, D.-G., Liu, W.-Y., & Chen, G.-T. (2013). A simple method for the isolation and purification of resveratrol from *Polygonum cuspidatum*. *Journal of Pharmaceutical Analysis*, *3*(4), 241–247. https://doi.org/10.1016/j.jpha.2012.12.001

Wang, Z., Zhao, X., Zhao, D., Zu, Y., Wang, L., Wu, M., Wang, L., Liu, Y., & Zhang, Q. (2021). Ultrasonic microwave-assisted micellar extraction and purification of flavonoids from licorice by metal complex and antisolvent recrystallization. *LWT*, *147*, 111501.

Wang, Y., Zhou, X., Wei, S., Wang, G., & Xi, J. (2023). Current status and future challenges in extraction, purification and identification of Cepharanthine (a potential drug against COVID-19). *Separation and Purification Technology*, *309*, 123038. https://doi.org/10.1016/j.seppur.2022.123038

Waseem, R., & Low, K. H. (2015). Advanced analytical techniques for the extraction and characterization of plant-derived essential oils by gas chromatography with mass spectrometry. *Journal of Separation Science*, *38*(3), 483–501.

World Health Organization. (2011). *Quality control methods for herbal materials* (rev. ed.). WHO Press. https://iris.who.int/bitstream/handle/10665/44479/9789241500739_eng.pdf

Wu, P., Sun, L., Qi, S., Zeng, T., Hou, Z., Yang, Y., Xu, L., Wei, J., & Xiao, P. (2023). Extraction process, chemical profile, and biological activity of aromatic oil from agarwood leaf (Aquilaria sinensis) by supercritical carbon dioxide extraction. *Journal of CO2 Utilization*, *77*, 102615. https://doi.org/10.1016/j.jcou.2023.102615

Zhu, J.-J., Yang, J.-J., Wu, G.-J., & Jiang, J.-G. (2020). Comparative antioxidant, anticancer and antimicrobial activities of essential oils from *Semen Platycladi* by different extraction methods. *Industrial Crops and Products*, *146*, 112206.

Zygler, A., Wasik, A., Kot-Wasik, A., & Namieśnik, J. (2012). The content of high-intensity sweeteners in different categories of foods available on the Polish market. *Food Additives & Contaminants: Part A*, *29*(9), 1391–1401.

4 Pulsed Electric Field Extraction of Essential Oils

Olawale P. Olatidoye

4.1 INTRODUCTION TO ESSENTIAL OILS AND EXTRACTION METHODS

Essential oils constitute complex mixtures of volatile organic compounds, typically found as condensed liquids within the living tissues of aromatic plants (leaves, flowers and seeds) and can be extracted in a concentrated form by physical methods (distillation and extraction, cold pressing or their combinations) which man have used due to their natural properties (fragrance and antibacterial/fungal properties) (Hassan and Hamad, 2009; Liu, 2009; Dreger and Wielgus, 2013; Rozalli, 2014). Essential oils are usually extracted from the plant tissues by different methods which include: vacuum distillation, steam distillation, supercritical CO_2 extraction and solvent extraction. Steam distillation is known to be the most commonly used in the industry because it does not require the use of organic solvents with high extraction yield. Extraction processes are determined by the mass transfer of target substances from within the cell through the cell wall. A major barrier to net movement of ions and molecules is a living or biotic membrane that separates the inner contents from the outside. The overcoming of cell wall can increase the extractability of intracellular compounds and improve the yield. The movement of essential oil through the cell wall by diffusion and osmosis is the most energy-intensive stage and determines the process speed.

Essential oils are valuable raw materials in various industries such as cosmetics, soap, toilet products, pharmaceuticals, confectionery food products, soft drinks, distilled alcoholic beverages, insecticides and fragrance industries, which represent nearly 60% of the total demand for natural substances and are characterized by a wide variety of products with relatively small quantities and often high prices (Shahi et al., 2009; Chauhan et al., 2009). Several factors have been alluded to the fact that oil content (yield) and composition and quality of the oil are important quality factors for oil. Some of these factors are place of origin, harvest season, weather, method of extraction and environmental conditions (Yajun et al., 2017). Due to economic and ecological factors, agri-allied industries are searching for alternatives to use new technologies to reduce CO_2 emissions and energy

DOI: 10.1201/9781032634180-4

consumption (Bousbia et al., 2009). In the past few years, several novel technologies have evolved or unfolded by researchers in the food industry to upgrade and/or to replace the existing methods as alternatives for highly competitive markets by reducing energy consumption. In this regard, the application of pulsed electric fields (PEFs) is a promising new technique. Moreover, several unconventional methods (microwave heating, ultrasound, PEF, supercritical fluids and ohmic heating) have been investigated to improve the overall yield and selectivity of bioactive components from plant material (Ghafoor et al., 2011; Toepfl et al., 2006; Lakkakula et al., 2004). Hydro-distillation is the traditional method with low oil yield, easy to operate and very economical. Solvent extraction was developed to overcome the problem of hydro-distillation with improvement in the yield of essential oil. However, solvent-free product is difficult to produce. Supercritical CO_2 extraction techniques have the disadvantage of high cost but the extraction of essential oils from different plants such as *Rosa damascena*, Eucalyptus, Thyme and Marrubium can be improved considerably using PEF (Miloudi et al., 2018a; Yajun et al., 2017; Tintchev et al., 2012). PEF can also be employed as a pretreatment method before distillation on biological cell with the formation of spores in the cell membrane through a process known as electroporation. This mechanism of extraction makes use of PEF which enhances the release of bioactive components and facilitates the process of diffusion in the tissues, thereby minimizing environmental risk by emitting less CO_2 into the atmosphere (Palai, 2015; Ferhat et al., 2006).

PEF has attracted progressive growth as a nonthermal technology in the food industry where it has been observed as an innovative nonthermal processing technology appropriate for the pretreatment of liquid and semi-solid edible products with low electrical conductivity (Barsotti and Cheftel, 1999; Bendicho et al., 2003). PEF has been reported to be suitable for essential oil extractions, due to the ability to liberate intracellular species with lower energy expenditures (Bendicho et al., 2002a). PEF techniques are used to optimize the energy consumption and preserve the qualitative properties of the food products used in biotechnology, pharmacy, making stimulus lasers. This process usually involves expanding the size of the cell membrane pores, thereby increasing its permeability, and facilitates the process of mass transport to the cell surface and forces the extraction of target substances. The food material is treated with a high voltage electric field so it can accumulate specific energy from 10 to 20 kJ/kg. Haji-Moradkhani et al. (2009) also reported that PEF application has the ability to increase the extraction efficiency with no significant effect on refractive index, acidity and peroxide value of oil. Abenoza et al. (2013) reported that PEF treatment could lead to a reduction of malaxation temperature from 26°C to 15°C without any negative flavor or taste production in the virgin olive oil. Essential oils are products of secondary metabolism, and PEF processing during essential oil extraction has shown enhanced essential oil yield of about 20–40% and has shortened the distillation process (Dunn, 1996; Everendilek et al., 1999). Several authors have reported that PEF has the benefit of speed, efficiency, low power and low pollution and they also involved in nonthermal pasteurization of food and extraction of value components from different natural sources (Dunn, 1996; Everendilek et al., 1999; Grahl and Markl, 1996).

Agri-food and chemical industries are facing the challenge of reducing CO_2 emissions and energy consumption and improving the overall yield and selectivity of bioactive components from plant material, hence the need to use new technology such as PEF (Bousbia et al., 2009). Therefore, the extraction yield from different plants can be improved considerably using PEF which is based on the electroporation of the cell membranes using high-level electrical field pulses, thereby releasing the intracellular content more easily which may facilitate the processes of diffusion in the tissues and the mass exchange with the external environment. Hence, the PEF technique will minimize environmental risks by emitting less CO_2 into the atmosphere (Ferhat et al., 2006). PEF is known to reduce energy consumption during mechanical operation and can enhance heat and mass transfer operation processes which are effectively deployed in drying and freezing of foods. PEF technology can improve the extraction efficiency by increasing the yield, decreasing the processing time and reducing the energy requirements of the process and subsequent processes. A number of researchers have employed PEF techniques for increasing the efficiency in essential oil extraction from some plants: 4 kV/cm was applied to *Rosa alba* L. with 30–250-minute extraction time by Dobreva et al. (2010) and to *Nepeta transcalcasica* with 120-minute extraction time by Dobreva et al. (2013); 0.8–2.5 kV/cm was applied to *Marribium vulgare* with 30–60-minute extraction time by Miloudi et al. (2018a); 10–30 kV/cm was applied to flowers of Damask rose with 30–180-minute extraction time by Zhou et al. (2017) and 1–3 kV was applied to *Artemisia herba alba* with 30–60-minute extraction time by Zhou et al. (2017) and Miloudi et al. (2018b), from which it was possible to extract higher essential oil quantities and to which the PEF parameters were optimized. Considering the potential of PEF, as a pretreatment for increasing the essential oil extraction from plants and reducing the distillation time, the number of studies published and plants investigated is still very limited (Dobreva et al., 2010; Miloudi et al., 2018a; Zhou et al., 2017). Hence, the aim of this chapter is to mark out the importance of PEF as a modern method of essential oil extraction and as an effective tool to extract essential oil from various raw materials and show the efficacy of PEF as a pretreatment method in extraction process. In addition, principles associated with mechanistic approaches, effect of treatment on food quality and problems associated with the consumer acceptability are emphasized in this chapter.

4.2 SOURCES OF NATURAL ESSENTIAL OILS

Natural essential oils are generally produced from one or more plant organs which can be classified as flowers: rose, jasmine, clove, mimosa, rosemary, lavender; leaves: *Ocimum* spp., lemongrass, jamrosa and mint; leaves and stems: geranium, verbena, cinnamon; bark: cinnamon, cassia, canella; wood: cedar, sandal, pine; roots; angelica, sassafras, saussurea, valerian; seeds: fennel, coriander, caraway, dill, nutmeg; fruits: orange, lemon, juniper; rhizomes: ginger,

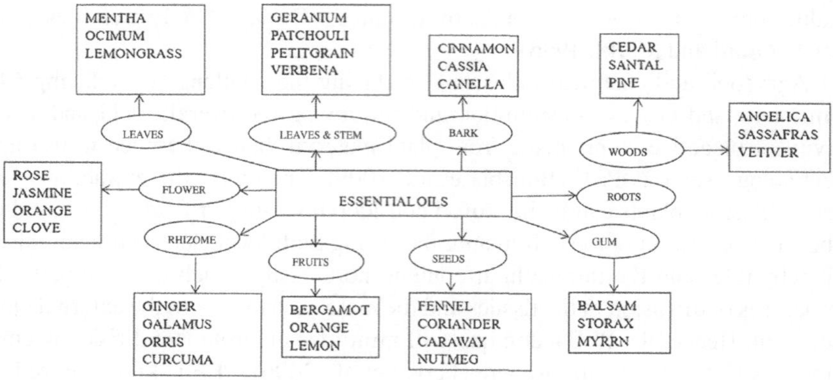

FIGURE 4.1 Essential oils present in plant parts.

Source: Created by the author.

curcuma, orris, and gums: balsam of Peru, balsam of Tolu, storax, myrrh, benzoin as shown in Figure 4.1. Specialized secretory plant structures may also produce and store essential oils as shown in Figure 4.1. These specialized plant structures includes glandular hairs, modified parenchymal cells, resin canals, oil tube, and lysigenous cavities which are dependent on plant family. In addition, the plant type and specific plant family affect the type of structure in which the oil is contained. However, the classification of such essential oils could be grouped into superficial and subcutaneous oils. Other factors that may affect both the quantity and quality of essential oil include humidity, wind, rain, surface temperature, harvesting schedule, and time of collection. These may also affect the perfumery values of oils.

4.3 CONSTITUENTS OF ESSENTIAL OIL

The major constituents of essential oils, as shown in Figure 4.2, consist of hydrocarbons, esters, terpenes, lactones, phenols, aldehydes, acids, alcohols and ketones. The oxygenated compounds (alcohols, esters, aldehydes, ketones, lactones, phenols) are the principal odor source among them. The understanding of the physical characteristics (boiling point, thermal stability and vapor–pressure–temperature relationship) and individual composition is of greatest importance in technological development of oxygenated compounds.

4.4 METHODS OF PRODUCING ESSENTIAL OILS

Essential oils perform several functions, including acting as insect repellents, aiding in pollination, serving as antimicrobial agents, and having a pleasant smell to humans. The methods of essential oil extractions are essentially carried out on a small or laboratory scale but some have found application in the industry on a

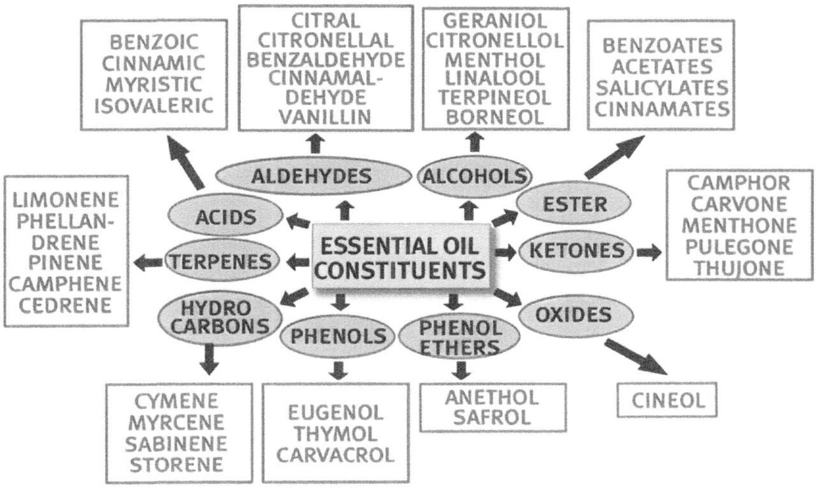

FIGURE 4.2 Chemical compounds present in essential oil.

large scale. The common forms of extraction used in essential oil industry include water distillation, water and steam distillation, and direct steam distillation, which exhibit same theoretical considerations of two-phase systems as shown in Figure 4.3. The main difference among these methods is found in the method of handling the material for extraction. Expression method is used for volatile oils which cannot be distilled without decomposition such as in lemon oil and orange oil. Hence, the peel must be removed from the fruit before the extraction process is carried out. Enfleurage is used on fresh plant parts with small quantity of volatile oil where the flower petals are spread on a fat for few hours to remove the essential oil and then followed by extraction with alcohol. This extraction method has found several applications in the production of perfumes and pomades. Moreover, most modern essential oil production makes use of volatile solvents, such as hexane and petroleum ether, which have the advantage of maintaining uniform temperature throughout the extraction process. Hence the essential oil produced has a more natural odor than the distillation method, which may have experienced some chemical modifications and alterations by the high temperature used. However, the distillation process presents a low-cost method.

4.5 IMPORTANCE OF ESSENTIAL OILS IN VARIOUS INDUSTRY

4.5.1 Anticancer Mechanism of Essential Oils

The constituents of natural essential oils play an important role in prevention and treatment of cancer due to their chemopreventive properties. These properties (antioxidant, antiproliferative, and antimutagenic) enhance detoxification and synergistic action of these constituents. There is a direct relationship between

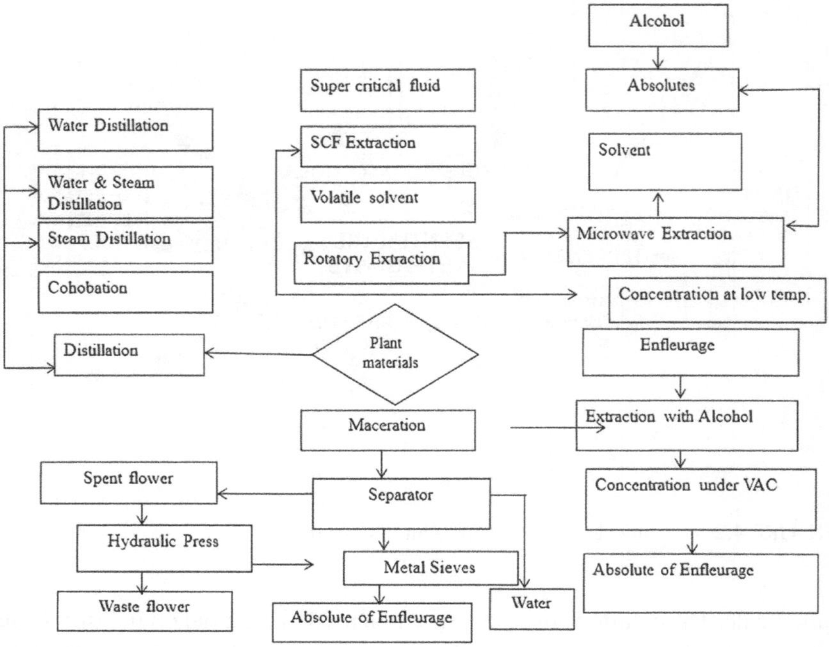

FIGURE 4.3 Production method for essential oil.

the production of reactive oxygen species (ROS) and the origin of oxidation and inflammation that can lead to cancer. Mitochondrial DNA damage can result from oxidative stress which leads to an increase in the mutation rate within cells and thus promoting oncogenic transformation. Storz (2005) reported that ROS activate signaling pathways which are known to encourage tumor development-regulating cellular proliferation. Arenas et al. (2011) also reported from their study that essential oils react with reactive oxygen to form reactive phenoxy radicals which can then react with ROS to prevent further damage. Essential oil can also bring about expression of antioxidant enzymes (catalase, superoxide dismutase, glutathione peroxidase and glutathione) with an increase in intracellular antioxidant activity which may lead to significant reduction in tumor mass. Figure 4.4 shows the preventive mechanisms of essential oils against various types of cancers.

4.5.2 Essential Oil as an Antioxidant Agent: Antioxidant Properties of Essential Oil

Several medicinal plants (basil, clove, cinnamon, nutmeg and thyme) have demonstrated the presence of essential oils with attested and authenticated radical scavenging and antioxidant properties due to the composition of the chemical constituents (Tomaino et al., 2015). Also, Butterfield and Lauderback (2020) reported that essential oils have a considerable high radical scavenging activity

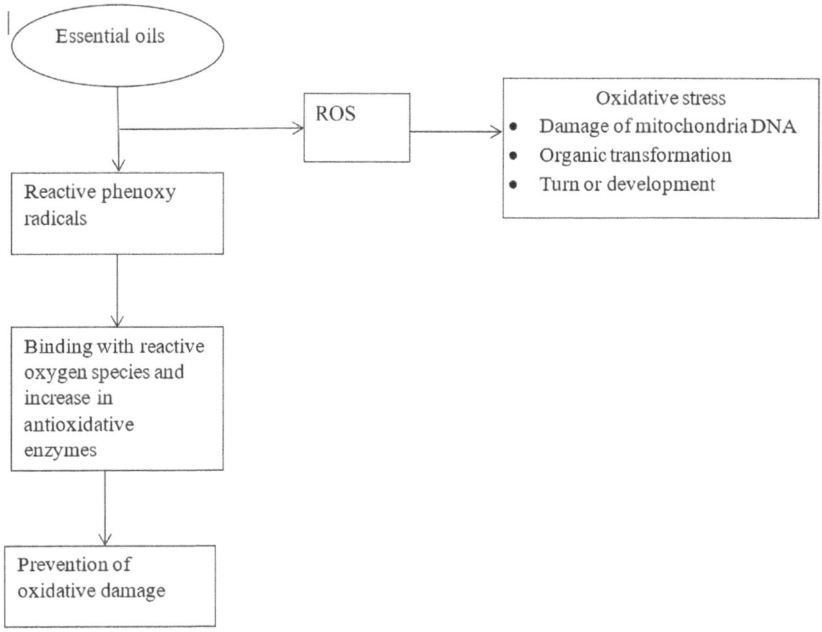

FIGURE 4.4 Chemopreventive mechanism of essential oil.

Source: Modified from Sani et al. (2012).

which is a result of ROS leading to the oxidation of biomolecules due to the alterations of molecules of chronic disorders native to cancer, diabetes, asthma, Parkinson's disease and aging. These findings are also supported by the report of the study carried out by Gardner (1997) on antioxidant properties of essential oils.

4.5.3 Essential Oil as an Antidiabetic Agent

Hyperglycemia is a condition associated with diabetes that arises as a result of the inability to either produce insulin or use it to regulate normal glucose levels in the blood. Inhibition of α-glucosidase and α-amylase is an important factor to control postprandial hyperglycemia in the management of type 2 diabetes mellitus as both enzymes are involved in the digestion of carbohydrates. α-amylase is involved in the breakdown of long-chain carbohydrates into disaccharides, while α-glucosidase breaks down starch and disaccharides to glucose or monosaccharides.

Thus, by inhibiting the enzyme, carbohydrate breakdown can be delayed, and ultimately the absorption of glucose in the bloodstream is reduced (Butterworth et al., 2011). Essential oils bind to the active site of the enzyme (α-amylase or α-glucosidase) and act as an inhibitor to form an enzyme–inhibitor complex thus inhibiting the enzyme activities (Figure 4.5). Several essential oils and their constituents have been analyzed for their antidiabetic potential such as essential

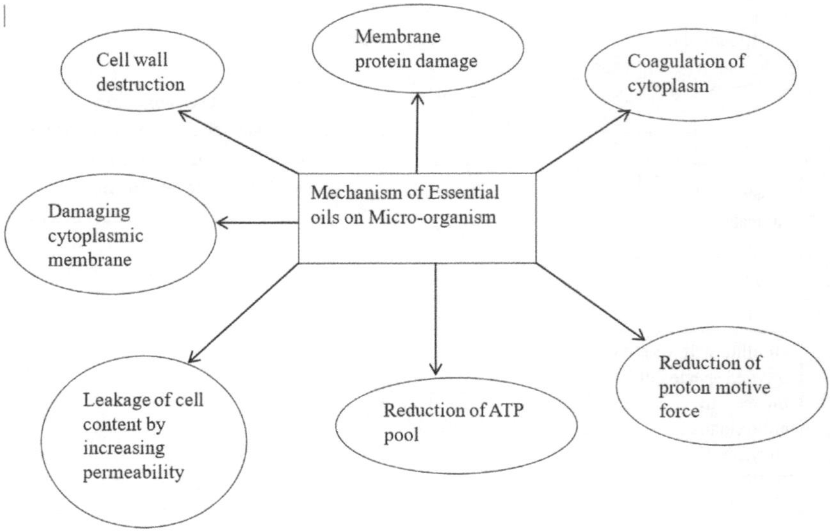

FIGURE 4.5 Essential oil action on microorganism.

oils of *Syzygium aromaticum*, *Cuminum cyminum* (Tahir et al., 2016), *Nepeta hindostana* (Joshi et al., 2021b), *Oliveria decumbens*, *Thymus kotschyanus*, *Trachyspermum ammi*, *Zataria multiflora* (Siahbalaei et al., 2020) and *Carthamus tinctorius* (Li et al., 2012).

4.5.4 Antimicrobial Application

Essential oils and their constituents play a vital role in antimicrobial activities. The antimicrobial activity of essential oil mainly depends on three characteristics – the nature of essential oil (hydrophilic or hydrophobic), its chemical constituents and the targeted organism (Fisher and Phillips, 2008; Holley and Patel, 2005). Due to their hydrophobic nature, essential oils pass across the cell wall and cytoplasmic membrane and disrupt the cell wall structure and make them more permeable. The membrane permeability leads to leakage of macromolecules and other cellular materials leading to cell death (Dorman and Deans, 2000). In bacteria, the permeabilization of the membranes is associated with loss of ions and reduction of membrane potential, the collapse of the proton pump and depletion of the ATP pool. Essential oils can also damage lipids and proteins by coagulating the cytoplasm. The antimicrobial activity of essential oils is mostly due to the presence of phenols, aldehydes and alcohols. Terpenoids are one of the major constituents present in essential oils and have oxygen atoms or methyl groups, which are localized or removed from specific enzymes by which they show greater activities. Generally, it has been observed that essential oils are more active in gram-positive bacteria than gram-negative bacteria due to the presence of peptidoglycan layer which lies outside the outer membrane. In contrast, in gram-negative bacteria, the outer membrane is composed of a double layer of phospholipids and is linked with the inner membrane by lipopolysaccharides; thus, hydrophobic macromolecules, such as essential oil constituents, are unable to penetrate the membrane which is responsible for the resistance of the gram-negative bacteria to essential oils. Aflatoxins, which are toxic secondary metabolites produced by common fungi, such as *Aspergillus flavus* and *A. parasiticus*, cause contamination of many food products. These aflatoxins are teratogenic, carcinogenic and mutagenic. Some essential oils not only inhibit the growth of such fungi but can also stop the production of aflatoxins. Essential oils are effective against a wide range of plants and human pathogenic bacteria, fungi and viruses by using different assays.

4.5.5 Pesticidal Applications of Essential Oils

Essential oils are distinctive mixtures of complex compounds that play a significant role in the plant's defense against bacteria, viruses, insects, fungi and herbs. Pesticides are chemical substances which have been reported to contain wide range of compounds displaying very different actions. Aktar et al. (2021) also reported that pesticides showed varying degree of different actions which include herbicides, insecticides, nematicides, rodenticides, avicides, algicides, fungicides and bactericides. Several disadvantages have been attributed to pesticides (chemical pesticides) in their actions due to factors such as high toxicity, environmental pollution and high cost. To this effect, new solutions have been put forward by several researchers with lower toxicity, fewer environmental damaging behavior and a better specific action while some botanicals are used for storage pests control due to the presence of the fumigant and their reaction/actions against a wide range of pests (Isman, 2000). Some oils with insecticidal and pesticidal actions appear to interact with the neuromodulator octopamine while others appear to interfere with GABA-gated chloride channels showing some neurotoxic mechanism of action. Essential oils have several applications in plant protection, such as insecticidal, herbicidal, nematicidal and fungicidal.

4.5.6 Essential Oil's Biological Activities

Several pharmaceutical and biological activities, such as antibacterial, antifungal, anticancer, antiviral, antidiabetic, antimutagenic, antiprotozoal, anti-inflammatory, antipyretic, analgesic, hepatoprotective, antidiarrheal, antihyperlipidemic, diuretic, neuroprotective and pesticidal activities, have been reported in various medicinal and aromatic plants containing essential oils (Figure 4.6) (Raut and Karuppayil, 2014). Across various plant families, essential oils are found to contain diverse therapeutic qualities. Specifically, their medicinal and antimicrobial properties are highly regarded for treating conditions such as constipation, dysentery, malaria, measles, stomach pain, yellow fever and aiding in dental care.

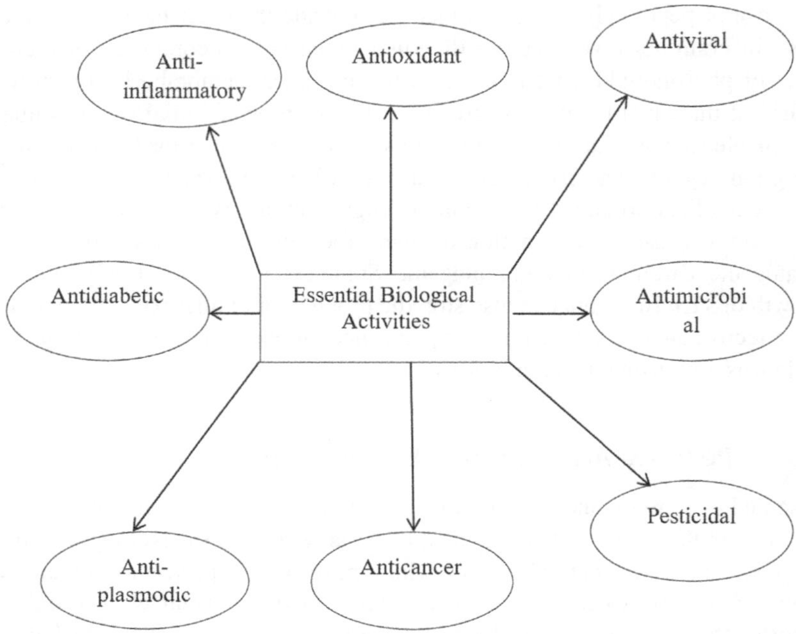

FIGURE 4.6 Essential oil's biological activities.

Source: Modified from Sonu Kumar Mahawer et al. (2022).

4.6 OVERVIEW OF TRADITIONAL EXTRACTION METHODS

The growing interest of the food, pharmaceutical and cosmetics industries in naturally occurring bioactive compounds or secondary plant metabolites also leads to a growing demand for the development of new and more effective analysis and isolation techniques. Traditional technologies used in essential oil processing are of great significance; some of the commonly used traditional methods include water distillation, water and steam distillation, steam distillation, cohobation, maceration and enfleurage. Maceration is usually adopted when oil yield from distillation is poor, while distillation methods are good for powdered plant materials, whereas solvent extraction is suitable for expensive, delicate and thermally unstable materials. Water distillation is the most commonly used method for the production of essential oil from plant materials.

4.6.1 The Need for Innovative Extraction Techniques

Essential oils which are present naturally in plants have found several applications by humans due to their therapeutic, cosmetic and nutritional benefits. This has made these desired compounds (essential oils) to play a significant role in drug and pharmaceutical industries. Extraction is the process of separating the desired component(s) from a chemical mixture, which is an important key step in using any bioactive natural compound. Technological advancement and research have led to the development of innovative techniques for the extraction and isolation of natural products leading to the better understanding of natural product chemistry and unlock new sources of valuable compounds. The first efforts by humans to make extractions from natural products probably coincided with the discovery of fire. Ancient cultures worldwide used water as an extraction medium during extraction processes. Later on, distillation and alcohol extraction methods were adopted and refined. The choice of solvent is the first important decision to be taken when designing an extraction procedure, as extraction solvents play a key role in determining the extracted compounds' quality, quantity and selectivity. The choice of extraction solvent depends on the chemical properties of the natural products being extracted and the desired end product. Solvents commonly used for natural product extraction include polar solvents, such as water, methanol, ethanol and acetone, and nonpolar solvents, such as hexane, chloroform and ethyl acetate. Each solvent has specific properties that make it suitable for extracting compounds of specific polarity. For example, ethanol–water mixtures are recommended for the extraction of phenolics. However, acetone has been proven efficient in the extraction of polyphenols from lychee flowers compared to methanol, water and ethanol. Solvents also play a role in the safety and potential environmental impact of the extraction process. It is important to consider the toxicity and flammability of solvents, as well as their potential effect on the environment. In this respect, organic solvents have a number of disadvantages: most of them are volatile and toxic and contribute significantly to environmental pollution. With the development of green chemistry, the design of green and sustainable extraction methods of natural products has become a hot research topic. Among the proposed green solvents, ionic liquids, natural deep eutectic solvents, supercritical and subcritical fluids and solvents from natural and renewable sources stand out as the most promising approaches for the current solvent innovation. The most common extraction method for natural products is typically solid–liquid extraction (the process of removing a solute component from the solid using a liquid solvent). Several solid–liquid extraction techniques have been applied to purify bioactive constituents from plants and microorganisms. Traditional methods include maceration, percolation and Soxhlet extraction. Recently developed extraction techniques for natural products are more energy-efficient and result in shorter extraction times compared to earlier methods. Such techniques include ultrasound-assisted extraction, microwave-assisted extraction and pressurized solvent extraction. The choice of extraction solvent and method depends on the type of natural raw material being extracted, the target compounds and the desired end product. Based on the law of similarity and intermiscibility (like dissolves like), solvents with a polarity value near the solute's polarity are likely to perform better and vice versa. When selecting the extraction method, it is necessary to take into consideration the stability of the desired product: temperature should be chosen carefully in order to avoid degradation. Contemporary methods for natural product extraction offer several benefits, including improved efficiency, reduced extraction time and reduced solvent usage. However, there are also some challenges: some methods, such as supercritical fluid extraction and microwave-assisted extraction, can be expensive due to the specialized equipment required to perform them. Given the costs involved, these same methods may also be challenging to scale up for commercial production. However, they also offer opportunities for increased productivity, potential selectivity and sustainability. Therefore, researchers must carefully evaluate these methods to determine their suitability for specific applications and weigh the benefits against the costs and limitations. Last but not least, interdisciplinary collaboration can contribute significantly to advancing this area of research. For example, a recent partnership of chemical engineers and biologists has led to clarification of the relations between extraction method and the antioxidant activity of C-phycocyanin. The interest in the extraction and isolation of natural products and their various applications is growing, with the discovery of new and more diverse applications for natural compounds driving the development of more efficient extraction methods. We hope this development will continue at a pace inspired by global need and directed by the environmental, safety and regulatory requirements such work demands.

4.6.2 Basic Principle of PEF-Assisted Extraction

Two views have been proposed by several authors concerning the potential of PEF mechanisms. The first view was shared by Min et al. (2003), who opined a PEF-based process on fast chemical reaction from several diverse compound so as to improve the quality of maximum concentration of solute that can dissolve in a solvent. The second view was put forward by Liang et al. (2002) and he opined that PEF technology was based on electroporation process involving an electric field produced by other charges that can improve which enhances the transmission of fluid into the cell membrane. The uncovering of cell membrane may lead to an increase in membrane voltage, resulting in the production of weak pores in the membrane. The mechanism of electroporation can be either reversible process or irreversible process, where the reversible electroporation requires 0.5 to 1.5kV/cm and irreversible electroporation requires 1.0 to 3.0 kV/cm, while values above 15kV/cm can cause microbial inactivation (Russell et al., 2000). The basic components of PEF processing system comprise a high-voltage pulse generator, a treatment chamber and a monitoring and controlling device. The food product is placed in the treatment chamber with either static design or continuous design with the two electrodes connected together with a nonconductive material in order to prevent or avoid electrical flow from one electrode to another. The cell membrane is punctured by creating hydrophilic pores which opened protein channels. This is followed by the production of electrical voltage pulses which are transmitted via the electrode where high-intensity electrical pulses are transmitted to the product. The membrane loses its structured functionality and the plant materials are extracted (Marquey et al., 1997). The food then experiences an electric field which is responsible for the irreversible cell membrane breakdown in microorganism (Mohamed and Amer Eissa, 2012). This is followed by the dielectric breakdown of the microbial cell membrane, and consequently, it interacts with the charged molecules within foods (Bendicho et al., 2003; Puértolas et al., 2004; Mohamed and Amer Eissa, 2012). The electroporation of cell and the formation of irreversible and reversible pores in their cell membranes were observed during the extraction process. The electroporation of cell brings about the development of irreversible and reversible pore formation in their cell membranes during the PEF extraction process which increases the extraction process. The strength of an electric field for delicate plant tissues typically ranges between 0.1 and 10 kV/cm, while tougher or harder materials require high intensities of 10 to 20 kV/cm for efficient extraction (Russell et al., 2000). This provides some beneficial effect such as nutrient preservation and liquid foods attributes such appearance, odor, flavour, taste and texture (Sale and Hamilton, 1997).

PEF techniques improve the transport of a substance (mass) by destroying the arrangement or organization of plasma membrane (Russell et al., 2000). Cell breakdown can be correspondingly increased due to permeability of body membrane and discharge of substances located within a cell leading to the increased rate of mass transfer. However, electric field is dependent on geometry, and electrode distance while electroporation intensity is also a function of type of pulse waveform, pulse waveform number, length of time and plant tissues (Qin et al., 1998). Formation of cavity or aperture (pore) is a constant process which is dependent on the electric field intensity, pulse length of time and pulse number. Permeation results from PEF-treated cells, leading to swelling and rupturing of the cell membrane.

4.6.3 Basic System Components of PEF Technology

The basic system components of PEF technology are composed of high-voltage power source, capacitor bank for energy storage, resistor for charging current and a switch for discharging energy from the capacitor to the food materials in the treatment chamber. Also, an oscilloscope is required for observing pulse wave form where high-voltage direct current (DC) generator then converts voltage into high DC then rectifies it into a high voltage DC. The power source is usually stored in the capacitor and is then discharged to generate an electric field in the treatment chamber to the food. The capacitor bank is charged by a DC power from amplified and rectified alternating current (AC). An electrical switch is used to discharge energy stored in the capacitor across the food materials in the treatment chamber. A pump is usually attached for a continuous system to transmit the food through the treatment chamber. Also, a cooling system could be attached to the chamber to reduce the effect of ohmic heating and also serve as temperature control during treatment. Probes are attached to measure voltage and current transmitted to the chamber. PEF systems for food processing consist of sample electrical system including a high-voltage repetitive pulser, treatment chamber, cooling system, voltage and current measuring device, a control unit and a data acquisition system. The treatment chamber is designed to hold food during PEF processing and house the discharging electrodes.

4.6.3.1 PEF Power Supply

High-voltage pulses are transmitted to the PEF system through a voltage pulse generator at the required field intensity, pulse wave form, pulse width, shape and duration. The high-voltage power supply for the system can be through DC or a capacitor power supply with high-frequency AC inputs (Zhang et al., 1996). Liquid foods are usually processed in a static treatment chamber or in continuous treatment chamber. Cold water from the cooling system is usually recirculated to prevent harmful thermal shock and to reduce heat generated by the electric current transmitted through the food (Barbosa-Cánovas et al., 1999).

4.6.3.2 PEF High-Power Capacitor

PEF technology high-power capacitor is mainly composed of storage capacitor and ON and OFF switches. Indicators may also be used due to high ohmic power consumption but play a less significant role. The energy generated is stored in capacitors which are then used to generate electric field. These electric fields are used to accelerate charged particles which may produce thermal, chemical, mechanical and electromagnetic waves. The magnetic fields produced bring about the production of high pressure from 0.1 GPa to several gigapascals. These effects are usually used to modify or compress, weld or destroy materials (Weise and Loeffler, 2001).

4.6.3.3 PEF Technology Switches

PEF switches are known to play an important function in the efficiency of the PEF system. The switches are known to be a discharging switch where the type of switch used will determine the performance efficiency, the quantity of current used and the voltage it can withstand. Suitable switches applicable for PEF systems are: spark gaps, trigatrons, thyratrons, semiconductors and ignitrons. However, Bartos (2000) reported that solid-state semiconductor switches are the future power switches. High-power switches are the connecting elements between the storage device and the load. The properties of the switches have been reported to be dependent on amplitude of pulse generator output, shape and time of rise in the pulse-forming element. Two main types of switches have been identified and reported by several authors which are: ON-only switches and ON/OFF switches. ON-only switches provide full discharge of the capacitor which can be turned off when discharge is completed. ON-only switches only accommodate high voltages with relatively lower cost. However, the main disadvantages of ON-only switches are their short life and low repetition rate. Common examples of ON-only switches include Ignitron, Gas Spark Gap, Trigatron and Thyratron. The ON/OFF switches have the advantage of longer lifespan and better performance than ON-only switches due to improvements made in semiconductor solid-state switches. Some examples of ON/OFF switches are gate turn-off thyristor, insulated gate bipolar transistor and symmetrical gate-commutated thyristor (Barbosa-Cánovas et al., 1999; Barsotti and Cheftel, 1999; Gongora-Nieto et al., 2002; Sepulveda and Barbosa-Cánovas, 2005).

4.6.3.4 PEF Technology in High-Voltage Pulse Generator

This generation of high-pulse voltage during PEF processing is essential to produce or deliver the desired electrical pulse voltage, geometrical figure and time during the formation of pulse network. The pulse-forming network is made up of several electrical circuit components, namely DC power supply, capacitors and resistors are linked together, a capacitor bank connected by two or more switches connected in parallel and pulse indicators and resistors. The capacitor bank is usually charged to the desired voltage by the DC power supply and provides reactive power for voltage support. The AC power is transformed into high-voltage AC power which is then converted from AC to DC (Xue and Zhang, 2017). The generalized PEF system consists of a high-voltage pulse generator which has been applied to treat spoiled grape juice samples. Ho and Mittal (2000) reported a detailed high-voltage pulse generator consisting of 30 kV DC high-voltage digital delay or pulse generator, treatment chamber and pumping and recording device.

4.6.3.5 PEF Technology Treatment Chamber

The treatment chamber is a major element of the PEF device where the electric field setup connects with the treatment media. Different types of treatment chamber for PEF processing techniques have been identified by several authors which include parallel plate treatment, coaxial cylinder treatment chamber and colinear treatment chamber (Huang and Wang, 2009; Jaeger et al., 2010). The design considerations for the PEF technology treatment chamber include electric field distribution uniformity, temperature increase and flow characteristics. The efficiency of the treatment for some media should be carried out in a long round tube treatment made up of circular or oval cross-sectional electrodes. This is usually carried out by submerging in a water with a dry PEF treatment in a direct contact with their product and electrodes so as to prevent cross-contamination. Dunn et al. (1987) opined that for most favorable arrangement of electric field consideration should be made for satisfactory and suitable electrode shapes and insulator which will lessen or minimize the breakdown of dielectric effect on food thereby preventing local high electric field strength. The dielectric breakdown of food is known to be deleterious as they bring about the destruction of food causing electrode damage of electrode and insulator surface damage and treatment chamber explosion due to increased pressure.

4.6.4 PEF Treatment Chamber Design

4.6.4.1 Parallel Plate Treatment Chamber

Parallel plate static design is a straightforward and easy batch system chamber suitable for laboratory scale. Parallel plate provides uniform distribution of electric field which lowers their application due to lower electrical resistance. Also, the electric field distribution is dependent on electrode's length and distance (Huang and Wang, 2009). A research work using PEF-modified in a 4-ml ultraviolet cuvette designed by Zbinden (2011). Foltz (2012) had designed the simplest chamber for cell lysis observation using silver chloride electrodes glued on view slides of glass microscope. A continuous flow of parallel chamber was designed by Goettel et al. (2013) using stainless steel electrodes.

4.6.4.2 Co-Field Treatment Chamber

The co-field treatment chamber is designed for a continuous system and finds application in food processing and is suitable for large-scale processing. The treatment chamber makes use of hollow high voltage and grounded electrode with a circular hole kept at a predetermined distance by an insulation and space. The product is pumped by drilling which may produce the electrical load of high-voltage discharge circuit (Meneses et al., 2011). The main advantage of co-field treatment chamber is that it has the ability to produce the same voltage as the voltage across the treatment chamber (Barbora-Cánovas et al., 1999). Also, Flisar et al. (2014) had also come up with a design of a co-field treatment chamber using stainless steel electrodes with insulation and a gap electrode of 15 mm. In a similar vein, Kempkes et al. (2011) designed a spectrum of commercial PEF chamber consisting of four treatment cells and pipe diameter of 1.5 cm from 10 to 100,000 L/hr. Co-field chambers with small electrode surface area and large electrode gap will provide a high load resistance. In addition, the inner diameter of the insulator should be slightly pinched in order to produce a more homogeneous electric field. The treatment chamber design has a major impact on the performance of PEF treatment which has effect on various process parameters.

4.6.5 Factors Affecting Pulsed Electric Technology

The factors that have a significant effect on PEF technology can be classified into processing factors, product factors and microbial factors.

4.6.5.1 Processing Factors

There are two major factors that can have a significant effect on the degree of destruction of microorganisms during PEF processing operation. These factors are pulse electric field strength and time of treatment. The characteristics of the cell and the medium type are dependent on specific electric field strength and specific treatment time. Qin et al. (1998) reported that increase in the electric field intensity may lead to a corresponding increase in the electric microbial inactivation which is also a function of number of pulses and pulse duration. Any increase in any of these factors may bring about a corresponding increase in microbial inactivation. The intensity of electric field applied is also a function of time of treatment where critical treatment time decreases with higher electric field intensity above the critical electric field. This is in agreement with the report of Bendicho et al. (2003) who reported that protease activity of *Bacillus subtilis*-inoculated milk decreased with the increased time of treatment, field strength and rate of pulse repetition. In addition, the method used in the application of electric field pulse treatment has a significant effect on the degree of microorganism inactivation. Jayaram et al. (1992) also reported that PEF electric treatment done at temperature between 50°C to 60°C had a significant effect on microorganism inactivation as a result of increase in electrical conductivity of the solution at higher temperature. This assertion was supported by the report of the study carried out by Marquez et al. (1997) who observed that higher temperature treatment could cause changes in cell membrane permeability and fluidity as a result of membrane disruption of the cell. Calderón-Miranda et al. (1999) also observed an increase in the electric field intensity of 30, 40 and 50 kV/cm with a corresponding progressive decrease in the Listeria inocula population with a two-and-half cycle reduction in the number of pulses applied. Several authors have also reported reversible and irreversible cell damage during PEF treatment such as decrease in *Salmonella typhimurium* population with increased pulse number and temperature of treatment (Liang et al., 2002, Ulmer et al., 2002)

4.6.5.2 PEF Media Condition

The efficiency of the PEF has been reported to be dependent on the conditions the media of PEF processing are subjected to. The conductivity of the PEF media is very crucial and equally dependent on temperature, chamber design, electrode material and cell type which is known to increase with increase in temperature. Also, a very low media conductivity is not desirable because the generation of pulse electric field is generally impaired while a very high media conductivity will not generate any electric field. Therefore, optimal treatment media should be given adequate consideration because of the aforementioned factors. Several authors have investigated the effects of media conditions on the efficiency of PEF treatment which are in agreement with the literature reports (Vega-Mercado et al., 1996; Álvarez and Heinz, 2007). In addition, Töpfl et al. (2007) reported the occurrence of air bubbles, cell agglomeration which brings about non-homogeneities of electric field leading to PEF treatment efficiency reduction.

4.6.5.3 Electric Field Strength

One of the main factors affecting PEF efficiency is electric field strength that is desirable by critical electric field strength needed for the breakdown of plasma membrane in this study (Zimmermann, 1996). The value of critical electric field strength is dependent on cell geometry and cell size. The applied voltage and the distance between electrodes are used to evaluate electric field strength. Several authors have investigated the electrical field strength required for nonthermal inactivation of yeast and bacteria (Heinz et al., 2001).

4.6.5.4 Food Product Factors

The major food product factors influencing PEF technology are conductivity, pH, ionic strength and type of food. Food with high electrical conductivity generates small electric field peaks in the treatment chamber, and hence such food cannot be used for PEF treatment (Barbora-Cánovas et al., 1999). The increase in the differences between the conductivity of a medium and microbial cytoplasm usually leads to decrease in the rate of microbial inactivation due to the increase in conductivity of food. This also leads to increase in flow of ionic substances across the membrane. Jayaram et al. (1992) reported that the application of equal input energy may bring about increased rate of microorganism inactivation with a reduction in conductivity. During electroporation, the types of microorganisms have a significant effect on ionic strength and pH of the medium which affect the cytoplasmic membrane. Hence, during PEF treatment the combination of pH, ionic strength and antimicrobial compound of food would be a veritable means to improve microorganism inactivation (Lu et al., 2001; Liang et al., 2002). Bendicho et al. (2003) also reported a positive result from microbial inactivation of enzymes by PEF.

4.6.5.5 Microbial Factors

The most critical microbial factors affecting PEF processing are the type of microorganism, concentration of microorganism and microbial growth stage in the food system. Gram-positive bacteria are more resistant to PEF than gram-negative bacteria while the sensitivity of yeast toward electric field processing is higher than bacteria as a result of the larger size possessed by yeast (Qin et al., 1995a). The reaction of microorganisms toward PEF system depends on the number and type of microorganisms in the food system (Russell et al., 2000). Logarithmic phase cells are generally more sensitive to stress than lag and stationary phase cells. PEF technology is one of the best processes for microorganism inactivation in acid-sensitive media because during cell division, the cell membrane is more sensitive or reactive to the applied field, but the process does not achieve a complete microbial cell disintegration as opined by Grahl and Markl (1996). However, PEF technology does not have the ability to inactivate or kill endospores but can bring about inactivation of bacteria (*E. coli*) and microbial lipases which are heat-resistant (Calderón-Miranda et al., 1999).

4.7 PEF-ASSISTED EXTRACTION

Essential oils are volatile and ethereal liquid oils obtained from the stem, stalk, and leaf of plants consisting of complex mixtures of volatile compounds with different odor and aroma.

The technology employed in essential oil extraction, either traditional or modern method, is very important as this will bring about an improvement in the quality and yield of essential oil produced. PEF technology requires a less rigorous electric field strength than an elevated or extreme level of electric field strength and has the ability to reduce negative impacts of traditional heating or extreme techniques (boiling and ultrasound-assisted or microwave-assisted extraction). This technique requires a less intense or less rigorous electric field strength (20 to 80 kV/cm) than an elevated or extreme level of electric field strength (100 to 300 V/cm), where the mode of extraction could be either in batch or continuous mode. The mechanism involved in PEF-assisted extraction (PEFAE) (electroporation or electro-permeabilization) makes use of external electric force, which is used to augment the cell membrane permeability (Panja, 2018). A specific food substance can be placed between the electrodes and a high-strength electric field voltage, which punctures the cell membrane by the formation of hydrophilic pores and the membrane where its physical functionality and the extraction take place (Redondo et al., 2018). There are basically two types of PEFAE: batch and continuous PEFAE. In batch PEFAE, the sample is first treated with a small amount of solvent between two electrodes and then the treated samples are removed from the pretreatment unit and stirred at different intensities with the help of a magnetic stirrer to check the solvent evaporation in order to increase the operational time due to low capacity of the system. In continuous PEFAE, the mixture of solvents is pumped into the treatment chamber by a peristaltic pump at a constant fluid velocity (Ranjha et al., 2021). This technique is getting popularized for the intensification of essential oil extraction (Miloudi et al., 2018a). Therefore, continuous PEFAE is mostly in use instead of the batch process. Several studies reported by authors revealed that PEF pretreatment could significantly enhance the extraction rate of essential oils. The high-voltage supply charges all the capacitors and breaks down the spark gap, leading to an abrupt voltage shock to the treatment chamber. Essential oils are intense, volatile aromatic compounds, which easily evaporate, that give plants their wonderful scents. Each of these complex precious liquids is extracted from a particular species of plant life. These oils have potent antimicrobial factors with wide range of therapeutic constituents; hence, the oil is often used for flavor, therapeutic and pleasant smell in a wide range of foods, medicines, cosmetics and personal care such as skin and hair, mouth cleanser, gums, teeth and general hygiene due to their unique chemical profile that provides different benefits,

4.7.1 Application of PEF in Food Processing

The use of PEF processing techniques in food products has been reported by several authors in the mild heat treatment to eliminate pathogens and extend shelf life of fruit juices, dairy products, soup, egg white and extraction of sugar (sugar beet), fermented brewer's yeast and reduction of wastewater. This processing technology has been successfully utilized with fruit juices with thick or thin flow properties and fruit with specific conductance such as oranges (Qin et al., 1998; Evrendilek et al., 2000). Also, fruit juices subjected to PEF processing treatment experience less color change and less browning due to conversion to furfural (Yeom et al., 2000). Also, liquid food products treated with PEF method retain their freshness and improve their textural attributes and bacteriological safety (Dunn and Pearlman, 1987). However, the application of PEF to food products is possible without bubbles, specific conductance, and smaller size of liquid food. Also, solid foods usually undergo a continuous process because they cannot be pumped; hence, PEF cannot be applied to solid food materials; it can also substitute for drying of food materials as preservation method, modification of enzymes and waste management for and improvement in metabolite extraction. It has also been used to enhance drying of plant tissue such as potato, coconut, carrot, mango and apple slices (Ade-Omowaye et al., 2000, 2001; Tedjo et al., 2002). PEF-treated food products exhibited reduced grinding energy, reduced elasticity modulus and softer tissue as reported in potato, apple and carrot (Lebovka et al., 2004). PEF offers application in the vegetative cells of microorganisms through pasteurization and retention of heat-sensitive bioactive compounds which bring about shelf-life extension and microbial safety (Teopfl et al. 2007; Sui et al. 2010; Guerrero-Beltrán et al., 2010).

4.7.2 Application of PEF in Drying

The effect of PEF technology on drying of several food products has been studied by several authors which revealed their cell structure resistance to diffusion and mass transfer rates (Barba et al. 2015). Alam et al. (2018) reported in their study that PEF-treated carrot and sliced parsnip exhibited a significant change in color and texture where the drying time was reduced by 28% and 21% when dried at 70°C and 60°C, respectively. Wiktor et al. (2016) also reported in their study that PEF-treated carrot samples experienced a reduction in drying time to about 8.2% with an increase in water diffusion coefficient to 16.7% with a higher lightness and redness when compared to untreated PEF samples. Ostermeier et al. (2018) also studied the effect of PEF pretreatment on drying characteristics of onion where they opined that increase in electric field strength led to a corresponding increase in cell disintegration with an improvement in the movement of moisture to the surface of the food product and a reduction in drying time to about 30%. Telfser and Gómez Galindo (2019) reported a reduced drying time for pretreated drying of basil (*Ocimum basilicum* L.) leaves by 57% air-drying, 33% vacuum-drying and 25% freeze-drying with PEF-treated samples exhibiting similar sensory scores for color and aroma from the sensory evaluation conducted. Ade-Omowaye et al. (2001) reported a similar finding during drying of carrots, potatoes, apples, coconuts or paprika where they reported that PEF accelerated the drying process of the food samples.

4.7.3 Application of PEF in Food Freezing

Application of freezing to food such as soft fruits and leafy vegetables has been reported to bring about loss of tissue due to shape loss and formation of ice crystals and thawing of freeze-dried food products. Demir et al. (2018) demonstrated that PEF-pretreated freeze-dried spinach leaves retained their turgor and viability after freezing and thawing cycles. Shayanfar et al. (2014) also reported that PEF-treated carrot and potato strips exhibited higher firmness, color and retained structural integrity after thawing than untreated samples, but the said PEF treatment alone was not a suitable pretreatment method. Jalté et al. (2009) demonstrated that PEF pretreatment can reduce freezing time, increase the freeze-drying rate and improve the quality of freeze-dried potatoes. In addition, Wiktor et al. (2015) demonstrated that PEF-pretreated freeze-dried apple tissues exhibited total freezing time reduction of 3.5–17.2% and total reduction in thawing time of 71.5%. These authors opined that multicellular tissue electroporation of the cell gave a reliable connection between intracellular and extracellular content allowing increased probability of ice nucleation and faster ice propagation after freezing and correspondingly shortening the freezing time.

4.7.4 PEF in Fruit and Vegetable Processing

The application of PEF technology in the fruit industry has been reported by several authors which is used in microorganism inactivation and production of off-flavors during storage (Hendrix and Redd, 1995). Jemai and Vorobiev (2002) also reported that PEF-treated apple juice augments or boosts the diffusion coefficient of soluble substances. The technology has been applied to enhance the removal of essential oil within the cell of several fruits by disruption of biological mass. PEF technology uses short pulses to enhance extraction of pectin by preventing excessive heat and undesirable electrolytic reactions (Min et al., 2007). PEF technology does not produce any change in pH, titratable acidity, color, Brix, conductivity, metal ions, non-enzymatic browning and vitamin concentration (Akin and Evrendilek, 2009). It has been reported that PEF-treated apple juice had better shelf life with no significant changes in physical, chemical or chemical in sugar and vitamin C. Zhang et al. (2018) also reported PEF-treated reconstituted orange juice exhibited improved shelf life at ambient temperature and also fruit and vegetable juices experienced improved sensory qualities of juices.

4.7.5 PEF-Treated Milk and Milk Products

PEF technology has been reported to be applicable in simulated milk ultrafiltration and skim milk has been effective up to 55–60°C with significant reduction in microbial load. In addition, Sharma et al. (2014), showed that whole milk had a 5–6log reduction in bacterial number at 22–28 kV at 50°C for 17–101 μs with different doses of PEF treatment to render the bacteria inactive. PEF technology treatment has shown a better rennetability property where an increase in electric field intensity led to significant increase in temperature and pulse number which also brings about a significant effect on properties of coagulated milk. PEF technology does not in any way affect milk particle sizes, the amount of protein in the serum and state of the whey protein which is a result of dissociation of the casein micelles due to PEF application. Hence, PEF technology possesses the potential and capability to bring about a microbiologically protected and guarded product and to preserve the bio-functionality of the milk fat globule membrane (Xu et al., 2015). In a similar study reported by Kumar et al. (2015), the PEF-treated raw milk at 40 kV/cm, 30 pulses and time of 2 μs was found to increase the shelf life by 14 days at 4°C. In addition, PEF technology can be used to pasteurize heat-sensitive emulsion products (salad dressing) without destroying the integrity of the emulsion. Similarly, food ingredients, sensitive foods and fermented food products pasteurized with these techniques lead to quality products' retention and superior quality products. Treatment of liquid milk can lead to shelf-life extension and reduced flavor degradation at 55°C, which may find application in the production of several dairy products with enhanced and better flavor with no change in any of the physicochemical and sensory attributes. Application of electric pulses to dairy protein and other food ingredients may bring about reduction in the spoilage microorganism without any visible change in functional and flavor properties.

4.7.6 PEF Meat and Poultry Products

PEF technology has been reported to have valuable contribution in its solid application with significant micro-structural changes or modification (improved tenderness, increase in purge loss and not cooking loss, changes in some quality parameters, bioactive compounds removal, but without affecting the polyunsaturated/saturated fatty acid ratio, omega a3/omega 6 ratio and free fatty acid profile). The technology has also found application in cell permeation enhancement to increase tenderness, indigenous microbiota attenuation to improve meat volatile profile maintenance during storage (Faridnia et al., 2015). PEF treatment on beef muscle at 5–10 kV with frequencies of 20, 50 and 90 Hz showed a reduction in shear force by 19% and enhanced softness and juiciness when stored for 21 days at refrigerated conditions. The use of PEF technology on egg products revealed the usefulness of the technology in making eggs safe for a long period of time.

4.8 LIMITATION/PROBLEMS AND CHALLENGES OF PET TECHNOLOGY

Several technical limitations of the PEF technology have been highlighted by several researchers which include the following:

1. The initial cost of installation is very high.
2. PEF technology does not have a significance on enzymes and spores.
3. A better processing condition can only be achieved if the particle size in the liquid medium is smaller in the gap in the treatment region.
4. The technology has limited application due to the fact that it is applicable to food product that can bear up against high electric field.
5. There is the limitation of nonuniform treatment, working and security problems as a result of globule of air or gas, or a globular vacuum formation during processing operation.
6. PET technology has limited application as it is confined or constricted to product that can withstand high electric field.
7. The energy efficiency of PEF technology has not been ascertained as its availability in commercial unit is still limited.
8. The major challenge for engineers is the production of equipment that are dependable on industrial scale.
9. The translation of the expert special skills into cheap and productive PEF-processing techniques within the national law and statute is still a serious challenge.

10. This electrochemical reaction that occurs may be detrimental to human being and difficult to replace; hence, the use of carbon electrodes and shorter pulses is an alternative.
11. Increasing the chamber's capacity becomes tricky, largely because the high-power systems needed for this expansion are designed without any leak current.

4.8.1 Future Directions

Researchers across the globe should come up with a project development design that gives a lot of attention to more effective comprehension of the technology of PEF product and processes relevant to execution at a profit-oriented level as the *outcomes* from the study are not good enough for total and outright abstraction of the effect involving food safety, and chemical parameters of the product and their processing conditions. There is a challenge of studying the effect of some processing conditions on safety of new products produced from PEF technology. However, its effect in combinations with other conventional processing technology on the nutritional, safety and quality parameters of fatty foods should be further studied. PEF technology has enormous and immense capacity to either replace or complement traditional food-processing techniques, but it is not certain if PEF technology as a new processing technology method will be adopted by food industry experts. Research on PEF technology should be geared toward microbial inactivation, reduction of spillage, alchemical reaction, holding or withholding of functional components and extraction of bioactive components in food during and after treatment with a detailed research study in order to explore the mechanism involved in these properties. Proper planning should be explored on the quality of being attractive to customers by providing accurate and simple information to consumer through suitable and appropriate labeling of products with a commodifying strategy. Research conducted showed that PEF can be used for the disordering or disarrangement of cells in food materials which can also be used in the treatment of wastewater and microorganism inactivation.

4.9 CONCLUSION

This chapter has shown that PEF technology is a new and original procedure of essential oil extraction so as to enhance volatile compounds' quality extracted from growing plant tissues by non-contact forces (electrostatic) or by electrically charged particles. The performance characteristics of PEF in this chapter centered on inactivation of microorganisms and disintegration of cells so as to further improve the quality mass transfer processes. This has brought about limitations as a result of the damage to sublethal microorganisms and ability of cells to recoup and reclaim stability or impregnability during chemical processes. PEF technology has the advantage of broad-based application in essential oil extraction which may lead to an increase in yield of oil extracted, but it may require a further study on composition of oil extracted. Hence, various research institutions should make efforts to develop aromatic plants for sustainable commercial exploitation. Concerted effort should be made to produce essential oil with consistent quality as there are variations in chemical compositions of essential oils. The side effects of temperature increase and electrochemical reactions must be evaluated in order to maintain the intended food quality. Also, adequate consideration should be given to optimization of the steps involved in the processing technology with the development of some innovations to reduce the cost of using PEF-processing techniques and to provide applications for a wide variety of foods. Also, adequate attention should be given to serious investigation of the effects of combining PEF with other preservation methods when upscaling equipment. Joint research among all the experts (food technologist, engineers, food chemist and microbiologist) is necessary for successful, safe and economical use (PEF) in food processing.

REFERENCES

Abenoza, M., Benito, M., Saldaña, G., Álvarez, I., Raso, J. (2013). Effects of pulsed electric field on yield extraction and quality of olive oil. *Food Bioprocess Technology*. 6:1367–1373

Ade-Omowaye, B. I. O., Angersbach, A., Eshtiaghi, M. N., Knorr, D. (2000). Impact of high-intensity electric field pulses on cell permeabilisation and as a pre-processing step in coconut processing. *Innovative Food Science & Emerging Technologies*. 1(3):203–209. https://doi.org/10.1016/S1466-8564(00)00014-X

Ade-Omowaye, B.I.O., Angersbach, A., Taiwo, K.A., Knorr, D. (2001). Use of pulsed electric field pre-treatment to improve dehydration characteristics of plant-based foods. *Trends in Food Science & Technology*. 12(8):285–295. https://doi.org/10.1016/S0924-2244(01)00095-4

Akin, E., Evrendilek, G.A. (2009). Effect of pulsed electric fields on physical, chemical, and microbiological properties of formulated carrot juice. *Food Science and Technology International*. 15(3):275–282. https://doi.org/10.1177/1082013209341414

Aktar, M.W., Sengupta, D., Chowdhury, A. (2021). Impact of pesticides use in agriculture: Their benefits and hazards. *Interdisciplinary Toxicology*. 2:1–12

Alam, M.D.R., Lyng, J.G., Frontuto, D., Marra, F., Cinquanta, L. (2018). Effect of pulsed electric field pretreatment on drying kinetics, color, and texture of parsnip and carrot. *Journal of Food Science*. 83(8):2159–2166. https://doi.org/10.1111/1750-3841.14216

Álvarez, I., Heinz, V. (2007). Hurdle technology and the preservation of food by pulsed electric fields. In *Food preservation by pulsed electric fields: From research to application*, Lelieveld, H.L.M., Notermans, S., de Haan, S.W.H., Eds.; Woodhead Publishing, pp. 165–177. https://doi.org/10.1533/9781845693831.2.165

Arenas, D.R.M., Acevedo, A.M., Méndez, L.Y.V., Kouznetsov, V.V. (2011). Scavenger activity evaluation of the clove bud essential oil (Eugenia caryophyllus) and eugenol derivatives employing ABTS+• decolorization. *Scientia Pharmaceutica*. 79(4):779

Barba, F.J., Parniakov, O., Pereira, S.A., Wiktor, A., Grimi, N., Boussetta, N., Saraiva, J.A., Raso, J., Martín-Belloso, O., Witrowa-Rajchert, D., Lebovka, N., Vorobiev, E. (2015). Current applications and new opportunities for the use of pulsed electric fields in food science and industry. *Food Research International*. 77:773–798. https://doi.org/10.1016/j.foodres.2015.09.015

Barbosa-Cánovas, G.V., Gongora-Nieto, M.M., Pothakamury, V.R., Swanson, B.G. (1999). *Preservation of foods with pulsed electric fields*. Academic Press Ltd., pp. 1–9, 76–107, 108–155

Barsotti, L., Cheftel, J.C. (1999). Food processing by pulsed electric fields: I. Physical aspects, II. Biological aspects. *Food Reviews International*. 15:163–213

Bartos, F.J. (2000). *Medium voltage AC drives*. Control Engineering Online, No. 20. http://www.controleng.com/archives/2000/ctl0201.00/000203.htm

Bendicho, S., Barbosa-Canovas, G.V., Martin, O. (2003). Reduction of protease activity in milk by continuous flow high intensity pulsed electric field treatments. *Journal of Dairy Science*. 86:697–703

Bendicho, S., Espachs, A., Arantegui, J., Martin, O. (2002a). Effects of high intensity pulsed electric field and heat treatments on vitamins of milk. *Journal of Dairy Research*. 69:113–123

Bousbia, N., Vian, M.A., Ferhat, M.A., Meklati, B.Y., Chemat, F. (2009). A new process for extraction of essential oil from Citrus peels: Microwave hydrodiffusion and gravity. *Journal of Food Engineering*. 90:409–413

Butterfield, D., Lauderback, C. (2020). Lipid peroxidation and protein oxidation in Alzheimer's disease brain: Potential causes and consequences involving amyloid beta-peptideassociated free radical oxidative stress. *Free Radical Biology & Medicine*. 32:1050–1060

Butterworth, P.J., Warren, F.J., Ellis, P.R. (2011). Human α-amylase and starch digestion: An interesting marriage. *Starch-Stärke*. 63(7):395–405

Calderón-Miranda, M.L., Barbosa-Cánovas, G.V., Swanson, B.G. (1999). Inactivation of *Listeria innocua* in skim milk by pulsed electric fields and nisin. *International Journal of Food Microbiology*. 51(1):19–30

Chauhan, R., Kaul, M., Shahi, A., Kumar, A., Ram, G., Tawa, A. (2009). Chemical composition of essential oils in Mentha spicata L. accession [IIIM (J) 26] from North-West Himalayan region, India. *Industrial Crops and Products*. 29:654–656

Demir, E., Dymek, K., Gómez Galindo, F. (2018). Technology allowing baby spinach leaves to acquire freezing tolerance. *Food and Bioprocess Technology*. 11(4):809–817. https://doi.org/10.1007/s11947-017-2044-7

Dobreva, A., Tintchev, F., Dzhurmansky, A., Toepfl, S. (2013). Effect of pulsed electric fields on distillation of essential oil crops. *Comptes rendus l'Académie Bulg. CR Academy of Bulgarian Science*. 66:1255–1260

Dobreva, A., Tintchev, F., Heinz, V., Schulz, H., Toepfl, S. (2010). Effect of pulsed electric fields (PEF) on oil yield and quality during distillation of white oil-bearing rose (Rosa alba L.). *Z. Fur Arznei Gewurzpflanzen*. 15:127–132

Dorman, H.D., Deans, S.G. (2000). Antimicrobial agents from plants: Antibacterial activity of plant volatile oils. *Journal of Applied Microbiology*. 88(2):308–316

Dreger, M., Wielgus, K. (2013). Application of essential oils as natural cosmetic preservatives. *Sciendo*. 59(4):142–156

Dunn, J. (1996). Pulsed light and pulsed electric field for foods and eggs. *Poultry Science*. 75:1133–1136

Dunn, J.E., Pearlman, J.S. (1987). *Methods and apparatus for extending the shelf life of fluid food products*. U.S. Patent No. 4,695,472, U.S. Patent and Trademark Office.

Evrendilek, G.A., Jin, Z.T., Ruhlman, K.T., Qiu, X., Zhang, Q.H., Richter, E.R. (2000). Microbial safety and shelf-life of apple juice and cider processed by bench and pilot scale PEF systems. *Innovative Food Science & Emerging Technologies*. 1(1):77–86. https://doi.org/10.1016/S1466-8564(99)00004-

Everendilek, G.A., Zhang, Q.H., Richter, E.R. (1999). Inactivation of Escherichia coli 0157:H7 and Escherichia coli 8739 in apple juice by pulsed electric fields. *Journal of Food Protection*. 62:793–796

Faridnia, F., Ma, Q., Bremer, P.J., Burritt, D.J., Hamid, N., Oey, I. (2015). Effect of freezing as pre-treatment prior to pulsed electric field processing on quality traits of beef muscles. *Innovative Food Science & Emerging Technologies*. 29:31–40. https://doi.org/10.1016/j.ifset.2014.09.007

Ferhat, M.A., Meklati, B.Y., Smadja, J., Chemat, F. (2006). An improved microwave Clevenger apparatus for distillation of essential oils from orange peel. *Journal of Chromatography A*. 1112:121–126

Fisher, K., Phillips, C. (2008). Potential antimicrobial uses of essential oils in food: Is citrus the answer? *Trends in Food Science and Technology*. 19(3):156–164

Flisar, K., Meglic, S.H., Morelj, J., Golob, J., Miklavcic, D. (2014). Testing a prototype pulse generator for a continuous flow system and its use for E. coli inactivation and microalgae lipid extraction. *Bioelectrochemistry*. 100:44–51

Foltz, G. (2012). *Algae lysis with pulsed electric fields*. Master Thesis, California State Polytechnic University, San Luis Obispo Follow

Gardner, P. (1997). Superoxide-driven aconitase FE-S center cycling. *Bioscience Reports*. 17:33–42

Ghafoor, K., Hui, T., Choi, Y.H. (2011). Optimization of ultrasonic-assisted extraction of total anthocyanins from grape peel using response surface methodology. *Journal of Food Biochemistry*. 35:735–746

Goettel, M., Eing, C., Gusbeth, C., Straessner, R., Frey, W. (2013). Pulsed electric field assisted extraction of intracellular valuables from microalgae. *Algae Research*. 2:401–408

Gongora-Nieto, M.M., Sepulveda, D., Pedrow, P., Barbosa-Cánovas, G., Swanson, B. (2002). Food processing by pulsed electric fields: Treatment delivery, inactivation level, and regulatory aspects. *Lebensm. Wiss. Technolnogy*. 35(5):375–388

Grahl, T., Markl, H. (1996). Killing of microorganisms by pulsed electric fields. *Applied Microbiology Biotechnology*. 45:148–157

Guerrero-Beltrán, J.Á., Sepulveda, D.R., Góngora-Nieto, M.M., Swanson, B.G., Barbosa-Cánovas, G.V. (2010). Milk thermization by pulsed electric fields (PEF) and electrically induced heat. *Journal of Food Engineering*. 100(1):56–60. https://doi.org/10.1016/j.jfoodeng.2010.03.027

Haji-Moradkhani, A., Rezaei, R., Moghimi, M. (2009). Optimization of pulsed electric field-assisted oil extraction from cannabis seeds. *Journal of Food Process Engineering*. 42:13028

Hassan, E.A, Hamad, E.H. (2009). The chemical constituent and vegetative and yielding characteristics of fennel plants treated with organic and bio-fertilizer instead of mineral fertilizer. *Australian Journal of Basic and Applied Sciences*. 3(2):579–587

Heinz, V., Álvarez, I., Angersbach, A., Knorr, D. (2001). Preservation of liquid foods by high intensity pulsed electric fields—Basic concepts for process design. *Trends in Food Science & Technology*. 12(3–4):103–111

Hendrix, C.M., Redd, J. B. (1995). Chemistry and technology of citrus juices and by-products. In *Production and packaging of non-carbonated fruit juices and fruit beverages*; Ashurst, P.R., Ed.; Springer, pp. 53–87. https://doi.org/10.1007/978-1-4615-0949-3_2

Ho, S.-Y., Mittal, G.S. (2000). High voltage pulsed electrical field for liquid food pasteurization. *Food Reviews International*. 16(4):395–434. https://doi.org/10.1081/FRI-100102317

Holley, R., Patel, D. (2005). Improvement in shelf-life and safety of perishable foods by plant essential oils and smoke antimicrobials. *Food Microbiology*. 22(4):273–292

Huang, K., Wang, J. (2009). Designs of pulsed electric fields treatment chambers for liquid foods pasteurization process: A review. *Journal of Food Engineering*. 95:227–239

Isman, M.B. (2000). Plant essential oils for pest and disease management. *Crop Protection*. 19(8–10):603–608. https://doi.org/10.1016/S0261-2194(00)00079-X

Jaeger, Hary, Menesen, Nicolas, Moritz, Jeldrik, Knorr, Duetrich. (2010). Model for the differentiation of temperature and electric field effects during thermal assisted PEF processing. *Journal of Food Engineering*. 100:109–118

Jalté, M., Lanoisellé, J.-L., Lebovka, N.I., Vorobiev, E. (2009). Freezing of potato tissue pre-treated by pulsed electric fields. *LWT – Food Science and Technology*. 42(3):576–580. https://doi.org/10.1016/j.lwt.2008.09.007

Jayaram, S., Castle, G.S.P., Margaritis, A. (1992). Kinetics of sterilization of *Lactobacillus brevis* cells by the application of high voltage pulses. *Biotechnology and Bioengineering*. 40(11):1412–1420

Jemai, A.B., Vorobiev, E. (2002). Effect of moderate electric field pulses on the diffusion coefficient of soluble substances from apple slices. *International Journal of Food Science & Technology*. 37(1):73–86. https://doi.org/10.1046/j.1365-2621.2002.00516.x

Joshi, M., Himani Kumar, R., Prakash, O., Pant, A.K., Rawat, D.S. (2021b). Chemical composition and biological activities of Nepeta hindostana (Roth) Haines, *Nepeta graciliflora* Benth. and *Nepeta cataria* L. from India. *Journal of Medicinal Herb*. 12(2):35–46

Kempkes, A.K., Roth, I., Gaudreau, M.P.J. (2011). *Pulsed electric field (PEF) method for continuous enhanced extraction of oil and lipids from small aquatic plants*. US Patent 2011/0107655 A1, 12 May 2011

Kineci, S. (2013). Influences of fermentation time, hydro-distillation time and fractions on Kumar, Yashwant, Patel, Krishna Kumar, Kumar, Vivek. (2015). Pulsed electric field processing in Food Technology. *Internal Journal of Engineering Studies & Technical Approach*. 1(2):6–17

Lakkakula, N.R., Lima, M., Walker, T. (2004). Rice bran stabilization and rice bran oil extraction using ohmic heating. *Bioresource Technology*. *92*(2):157–161. https://doi.org/10.1016/j.biortech.2003.08.010

Lebovka, N.I., Shynkaryk, N.V., Vorobiev, E. (2004). Effect of moderate thermal and pulsed electric field treatments on textural properties of carrots, potatoes and apples. *Innovative Food Science & Emerging Technologies*. 5(1):9–16. https://doi.org/10.1016/j.ifset.2003.12.001

Li, L., Wang, Q., Yang, Y., Wu, G., Xin, X., Aisa, H.A. (2012). Chemical components and antidiabetic activity of essential oils obtained by hydro-distillation and three solvent extraction methods from Carthamus tinctorius L. *Acta Chromatographica*. 24(4):653–665

Liu, K. (2009). Bioresource technology effects of particle size distribution, compositional and color properties of ground corn on quality of distillers dried grains with solubles (DDGS). *Bioresource Technology*. 100(19):4433–4440

Liang, Z., Mittal, G.S., Griffiths, M.W. (2002). Inactivation of Salmonella Typhimurium in orange juice containing antimicrobial agents by pulsed electric field. *Journal of Food Protection*. 65:1081–1087

Lu, J., Mittal, G.S., Griffiths, M.W. (2001). Reduction in levels of *Escherichia coli* O157:H7 in apple cider by pulsed electric fields. *Journal of Food Protection*. 64(7):964–969. https://doi.org/10.4315/0362-028X-64.7.964

Mahawer, Sonu Kumar, Himani, Sushila Arya, Kumar, Ravendra, Prakash, Om. (2022). Extractions methods and biological applications of essential oils. In *Essential Oils – Advances in Extractions and Biological Applications*. IntechOpen. http://dx.doi.org/10.5772/intechopen.102955

Marquez, V.O., Mittal, G.S., Griffths, M.W. (1997). Destruction and inhibition of bacterial spores by high voltage pulsed electric fields. *Journal of Food Science*. 62:399–409

Meneses, N., Jaeger, H., Knorr, D. (2011). Minimization of thermal impact by application of electrode cooling in a co-linear PEF treatment chamber. *Journal of Food Science*. 76(8):E536–E543. https://doi.org/10.1111/j.1750-3841.2011.02368.x

Miloudi, K., Hamimed, A., Benmimoun, Y., Bellebna, Y., Taibi, A., Tilmatine, A. (2018a). Intensification of essential oil extraction of the Marrubium vulgare using pulsed electric field. *Journal of Essential Oil Bearing Plants*. 21(3): 811–824 [CrossRef]

Miloudi, K., Tilmatine, A., Benmimoun, Y., Hamimed, A., Taibi, A., Bellebna, Y. (2018b). Intensification of essential oil extraction of Artemisia herba alba using pulsed electric field. In *Proceedings of the International Conference on Electrical Sciences and Technologies in Maghreb (CISTEM)*, Algiers, Algeria, 28–31 October 2018, pp. 1–5. [CrossRef]

Min, S., Jin, Z.T., Zhang, Q.H. (2003). Commercial scale pulsed electric field processing of tomato juice. *Journal of Agricultural and Food Chemistry*. 51:3338–3343

Min, S., Evrendilek, G.A., Zhang, Q.H. (2007). Pulsed electric fields: Processing system, microbial and enzyme inhibition, and shelf life extension of foods. *IEEE Transactions on Plasma Science*. 35(1):59–73. https://doi.org/10.1109/TPS.2006.889290

Mohamed, M.E.A., Amer Eissa, A.H. (2012). Pulsed electric fields for food processing technology. In *Structure and Function of Food Engineering*. http://dx.doi.org/10.5772/48678

Ostermeier, R., Giersemehl, P., Siemer, C., Töpfl, S., Jäger, H. (2018). Influence of pulsed electric field (PEF) pre-treatment on the convective drying kinetics of onions. *Journal of Food Engineering*. 237:110–117. https://doi.org/10.1016/j.jfoodeng.2018.05.010

Palai, A. (2015). *Analysis of essential oil content obtained from plant materials using steam distillation* (Doctoral dissertation). Ethesis

Panja, P. (2018). Green extraction methods of food polyphenols from vegetable materials. *Current Opinion in Food Science*. 23:173–182

Puértolas, E., Koubaa, M., Barba, F.J. (2016). An overview of the impact of electrotechnologies for the recovery of oil and high-value compounds from vegetable oil industry: Energy and economic cost implications. *Food Research International*. 80:19–26

Qin, B.L., Barbona-Canovas, G.V., Swanson, B.G., Pedrow, P.D. (1998). Inactivating microorganism using a pulsed electric field-continuous treatment system. *IEEE Transactions on Industry Application*. 34:43–49

Qin, B.L., Chand, F.J., Burhora-Canovas, G.V., Swanson, B.G. (1995). Non-thermal inactivation of S.cerevesae in apple juice pulsed electric fields. *Lebenson wais Technology*. 28: 564–568

Qin, B.L., Zhang, Q., Barbosa-Cánovas, G., Swanson, B., Pedrow, P.D. (1995). Pulsed electric field treatment chamber design for liquid food pasteurization using finite element method. *Transactions of the ASAE*. 38(2):557–565

Ranjha, M.M., Kanwal, R., Shafique, B., Arshad, R.N., Irfan, S., Kieliszek, M. (2021). A critical review on pulsed electric field: A novel technology for the extraction of phytoconstituents. *Molecules*. 26(16):4893

Raut, J.S., Karuppayil, S.M. (2014). A status review on the medicinal properties of essential oils. *Industrial Crops and Products*. 62:250–264

Redondo, D., Venturini, M.E., Luengo, E., Raso, J., Arias, E. (2018). Pulsed electric fields as a green technology for the extraction of bioactive compounds from thinned peach by-products. *Innovative Food Science & Emerging Technologies*. 45:335–343

Rozalli, M. (2014). Particle size distribution of natural peanut butter and its dynamic rheological properties. *International Journal of Food Properties*. 18(9):1888–1894

Russell, N.J., Colley, M., Simpson, R.K., Trivett, A.J., Evans, R.I. (2000). Mechanism of action of pulsed high electric field (PHEF) on the membranes of food-poisoning bacteria is an "all-or-nothing" effect. *International Journal of Food Microbiology*. 55(1–3):133–136. https://doi.org/10.1016/S0168-1605(00)00169-0

Sale, A.J.H., Hamilton, W.A. (1997). Effects of high electric fields on microorganisms I, Killing of bacteria and yeast. *Biochimica et Biophysica Acta*. 148:781–788

Sani, D.H., Munna, A.N., Alam, M.J., Salim, M., Alam, M. (2012). Evaluation of α-amylase inhibition and cytotoxic activities of the Arachis hypogaea and Cinnamomum tamala. *Current Nutrition and Food Science*. 17(3):328–336

Sepulveda, D.R., Barbosa-Cánovas, G.V. (2005). Present status and the future of PEF technology. In *Novel food process technologies*; Barbosa-Cánovas, G.V., Tapia, M.S., Cano, M.P., Eds.; CRC Press: Boca Raton, FL, pp. 1–45

Shahi, C., Leitch, M., Laforest, S. (2009). Marketing intelligence system for small-scale essential oils industry of North-Western Ontario. In *Seeing the forest beyond the trees: IUFRO 3.08 small-scale forestry symposium proceedings*, Morgantown, WV, June 7–11, 2009, pp. 227–236.

Sharma, P., Bremer, P.J., Oey, I., Everett, D.W. (2014). Bacterial inactivation in whole milk using pulsed electric field processing. *International Dairy Journal*. 35(2):49–56. https://doi.org/10.1016/j.idairyj.2013.10.005

Shayanfar, S., Chauhan, O.P., Toepfl, S., Heinz, V. (2014). Pulsed electric field treatment prior to freezing carrot discs significantly maintains their initial quality parameters after thawing. *International Journal of Food Science & Technology*. 49(4):1224–1230. https://doi.org/10.1111/ijfs.12421

Siahbalaei, R., Kavoosi, G., Shakeri, R. (2020). In vitro antioxidant and antidiabetic activity of essential oils encapsulated in gelatin-pectin particles against sugar, lipid and protein oxidation and amylase and glucosidase activity. *Food Science & Nutrition*. 8(12):6457–6466

Storz, P. (2005). Reactive oxygen species in tumor progression. *Frontiers in Bioscience*. 10(1–3):1881–1896

Sui, Q., Roginski, H., Williams, R.P.W., Versteeg, C., Wan, J. (2010). Effect of pulsed electric field and thermal treatment on the physicochemical properties of lactoferrin with different iron saturation levels. *International Dairy Journal*. 20(10):707–714. https://doi.org/10.1016/j.idairyj.2010.03.013

Tahir, H.U., Sarfraz, R.A., Ashraf, A., Adil, S. (2016). Chemical composition and antidiabetic activity of essential oils obtained from two spices (Syzygium aromaticum and Cuminum cyminum). *International Journal of Food Properties*. 19(10): 2156–2164

Tedjo, W., Taiwo, K.A., Eshtiaghi, M.N., Knorr, D. (2002). Comparison of pretreatment methods on water and solid diffusion kinetics of osmotically dehydrated mangoes. *Journal of Food Engineering*. 53(2):133–142. https://doi.org/10.1016/S0260-8774(01)00149-2

Telfser, A., Gómez Galindo, F. (2019). Effect of reversible permeabilization in combination with different drying methods on the structure and sensorial quality of dried basil (*Ocimum basilicum* L.) leaves. *LWT – Food Science and Technology*. 99:148–155. https://doi.org/10.1016/j.lwt.2018.09.062

Tintchev, F., Dobreva, A., Schulz, H., Toepfl, S. (2012). Effect of pulsed electric fields on yield and chemical composition of rose oil (Rosa damascena Mill.). *Journal of Essential Oil Bearing Plants*. 15:876–884

Toepfl, S., Heinz, V., Knorr, D. (2007). High-intensity pulsed electric fields applied for food preservation. *Chemical Engineering and Processing: Process Intensification*. 46(6):537–546. https://doi.org/10.1016/j.cep.2006.07.011

Toepfl, S., Mathys, A., Heinz, V., Knorr, D. (2006). Potential of high hydrostatic pressure and pulsed electric fields for energy efficient and environmentally friendly food processing. *Food Reviews International*. 22:405–423

Tomaino, A., Cimino, F., Zimbalatti, V., Venuti, V., Sulfaro, V., De, P.A. (2015). Influence of heating on antioxidant activity and the chemical composition of some spice essential oils. *Food Chemistry*. 89:549–554

Töpfl, S., Heinz, V., Knorr, D. (2007). High intensity pulsed electric fields applied for food preservation. *Chemical Engineering and Processing: Process Intensification*. 46(6):537–546. https://doi.org/10.1016/j.cep.2006.07.011

Ulmer, H.M., Heinz, V., Ganzle, M.G., Knorr, D., Vogel, R.F. (2002). Effects of pulsed electric fields on inactivation and metabolic activity of Lactobacillus plantarum in model beer. *Journal of Applied Microbiology*. 93:326–335

Vega-Mercado, H., Pothakamury, U.R., Chang, F.J., Zhang, Q., Barbosa-Cánovas, G.V., Swanson, B.G. (1996). Inactivation of *Escherichia coli* O157:H7 by combining pH, ionic strength, and pulsed electric fields hurdles. *Food Research International*. 29(2):117–121. https://doi.org/10.1016/0963-9969(96)00015-4

Weise, T.H.G.G., Loeffler, M.J. (2001). Overview on pulsed power applications. In *Proceedings of the International Conference on Pulsed Power Applications*, Gelsenkirchen, Germany, pp. A.0I/I-A.01/8

Wiktor, A., Schulz, M., Voigt, E., Witrowa-Rajchert, D., Knorr, D. (2015). The effect of pulsed electric field treatment on immersion freezing, thawing and selected properties of apple tissue. *Journal of Food Engineering*. 146:8–16. https://doi.org/10.1016/j.jfoodeng.2014.08.013

Wiktor, A., Nowacka, M., Dadan, M., Rybak, K., Łojkowski, W., Chudoba, T., Witrowa-Rajchert, D. (2016). The effect of pulsed electric field on drying kinetics, color, and microstructure of carrot. *Drying Technology*. 34(11):1286–1296. https://doi.org/10.1080/07373937.2015.1105813

Xu, S., Walkling-Ribeiro, M., Griffiths, M.W., Corredig, M. (2015). Pulsed electric field processing preserves the antiproliferative activity of the milk fat globule membrane on colon carcinoma cells. *Journal of Dairy Science*. 98(5):2867–2874. https://doi.org/10.3168/jds.2014-8839

Xue, Y., Zhang, X.-P. (2017). Reactive power and AC voltage control of LCC HVDC system with controllable capacitors. *IEEE Transactions on Power Systems*. 32(1):753–764.

Yajun, Z., Changmei, X., Susu, Z., Guangming, Y., Ling, Z., Shujie, W. (2017). Effects of high intensity pulsed electric fields on yield and chemical composition of rose essential oil. *International Journal of Agricultural and Biological Engineering*. 10:295–301

Yeom, H.W., Streaker, C.B., Zhang, Q.H., Min, D.B. (2000). Effects of pulsed electric fields on the quality of orange juice and comparison with heat pasteurization. *Journal of Agricultural and Food Chemistry*. 48(10): 4597–4605. https://doi.org/10.1021/jf000306p

Zbinden, M.D.A. (2011). *Pulsed electric field (PEF) as an intensification pretreatment for greener solvent lipid extraction from microalgae*. Master Thesis, University of Kansas, United State

Zhang, Q., Qin, B.-L., Barbosa-Cánovas, G.V., Swanson, B.G., Pedrow, P.D. (1996). *Batch mode food treatment using pulsed electric fields*. US Patent No. 5,549,041). U.S. Patent and Trademark Office.

Zhang, Q.W., Lin, L.G., Ye, W.C. (2018). Techniques for extraction and isolation of natural products: A comprehensive review. *Chinese Medicine*. 13:20. https://doi.org/10.1186/s13020-018-0177-x

Zhou, Y.J., Xue, C.M., Zhang, S.S., Yao, G.M., Zhang, L., Wang, S.J. (2017). Effects of high intensity pulsed electric fields on yield and chemical composition of rose essential oil. *International Journal of Agricultural and Biological Engineering*. 10:295–301

Zimmermann, U. (1996). The effect of high intensity electric field pulses on eukaryotic cell membranes: Fundamentals and applications. In *Electromanipulation of cells*; Zimmermann, U., Neil, G.A., Eds.; CRC Press, pp. 1–106.

5 Mechanochemically Assisted Extraction of Essential Oil

Jacky S. Bouanga Boudiomboa, Banele Vatsha and Olalere Olusegun Abayomi

5.1 INTRODUCTION

The extraction of naturally valuable substances such as polysaccharides (Zhu et al. 2012), flavonoids (Xie et al. 2013), phenolics (Zhang et al. 2020), polyphenols (Dobroslavić et al. 2022), and proteins (Liu and Xi 2021) from plants or other materials is highly in demand. These bioactive compounds have growing interest for promoting sustainable ingredients for modern food and pharmaceutical industries. There is a plethora of conventional extraction methods such as maceration, decoction, infusion, percolation, heat-reflux extraction and Soxhlet extraction (Gligor et al. 2019; Fan et al. 2022). Even though these extraction techniques are deemed relatively easy to use, they usually utilize toxic solvents and give poor extraction efficiency. In addition to this, there is also an extended extraction time and elevated extraction temperatures. This may lead to the degradation of thermolabile compounds and the decrease of the target compounds' pharmacological activity. Thus, a handful of "green" extraction methods have been researched, including ultrasound-, microwave-, pulse electric field-, and mechanochemical-assisted extraction (MCAE) (Aparecida de Marco et al. 2018). Figure 5.1 presents a broad view of MCAE method with some important parameters. It is worthy to note that "Mechanochemistry" has been recently listed as one of the chemical innovations to change our world.

MCAE is a speedy method that applies mechanical and chemical processes (Xie et al. 2011b). The mechanical processes entail the physical change of different substances by applying energy in the systems (Mateti et al. 2021a). On the other hand, the chemical process focuses on the chemical transformation of the materials, namely the chemical composition and chemical reactivity. Gathering these different properties, MCAE processes involve the changes in the material's structure, chemical composition, as well as chemical activity (Xie et al. 2013). One of the consequences of these changes lead to an increase in the solubility of some components, making its use as a pre-treatment method quite extensive (P. Wang and Tang 2009).

DOI: 10.1201/9781032634180-5

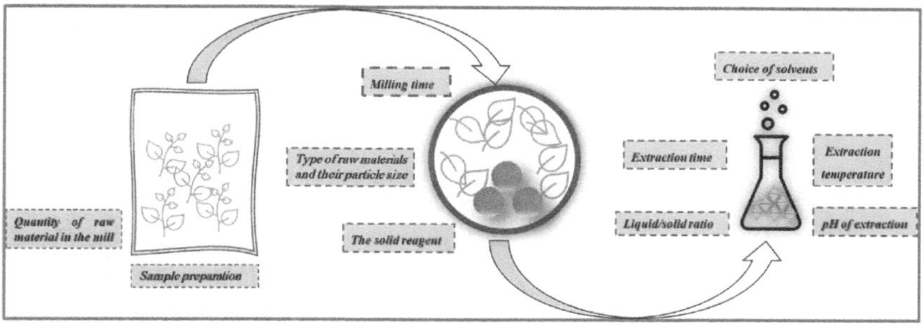

FIGURE 5.1 Adjustable parameters to improve extraction efficiency.

MCAE is an interdisciplinary science focused on the transformation of substances in all states of aggregation induced by mechanical energy (Xie et al. 2013). In general, the mechanical forces obtained by shear, friction, and impact will be applied on a substance (P. Zhu et al. 2022; Takacs 2013). The mechanical energy resulting from these mechanical forces will be converted to the substance's eternal energy. The above will then lead to changes within the material's where the structure, physical and chemical properties can be altered (Bychkov et al. 2019). The later properties are proven by the increase in the substance's surface area and the decrease in crystallinity and production of active free radicals (P. Zhu et al. 2022). Nevertheless, to understand the base of MCAE, the basic of mechanochemistry should be explored.

5.2 HISTORY OF MECHANOCHEMISTRY

MCAE falls under the umbrella of mechanochemistry. There are different sources of activation of chemical reaction, namely heat, irradiation, mechanical, and electrochemistry (L Takacs 2012). The mechanical activation is obtained when a reaction is activated by the direct absorption of mechanical energy (Ribas-Arino and Marx 2012). In this case, the energy input is obtained from mechanical forces. Therefore, mechanochemistry is a branch of chemistry that focuses on the physical and chemical changes of substances induced by grinding, shearing, or sliding. Although not archive, mechanochemistry has been employed inadvertently for more than 2,000 years. History has found that several households have employed mechanochemistry to reduce the particles sizes of several condiments to prepare various types of flours. Additionally, the multidisciplinary applications of mechanochemistry have increased its slow development to a major area of science (Boldyrev and Tkáčová n.d.). In general, mechanochemistry finds application in mineralogy, inorganic and organic synthesis, and tribology (Takacs 2013). These disciplines used mechanochemistry as a method but with independent nomination of the strategy employed. It was only later during the years that similarities were discovered and a name pioneered to the strategy of physical and chemical

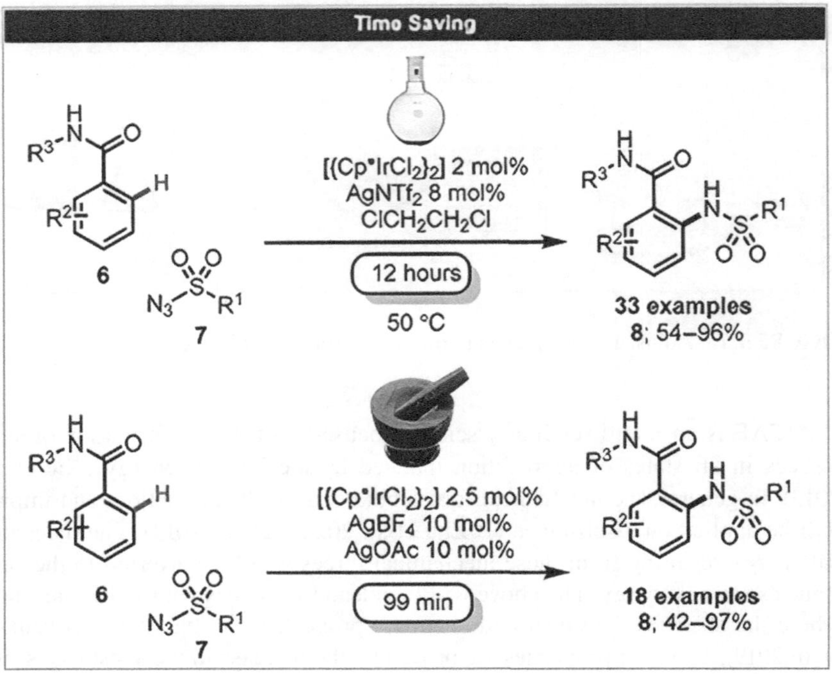

FIGURE 5.2 Mechanochemistry for the formation of C-H amidation of benzamides by Bolm and collaborators.

Source: Hermann et al. (n.d.) and Howard et al. (2018).

changes. This method has many advantages, one of them is to save time during the synthesis of many compounds (see Figure 5.2) (Howard et al. 2018).

Screening the book by Theophrasus (315 BC) where he described the formation of mercury using cinnabar in a mortar and pestle made from copper show a motif of mechanochemistry in history (Liu et al. 2022; Takacs 2000). Conversely, documentation for the direct application of mechanical stimulus was later recorded by Ling and Baker when they prepared halogen quinhydrone derivatives. However, the discovery of the identification of the various implication of the changes resulting from mechanical action was later observed by two main thespians: Walthere Spring and M. Carey Lea (Takacs 2003, 2018a; Boldyrev and Tkáčová n.d.).

Spring was responsible for the application of mechanochemistry in geology (Takacs 2013). At the beginning, Spring was looking for the base of the creation of minerals inside the earth and the chemistry behind it (Takacs 2018a). Nourished from this curiosity, he investigated the pressure on phase transformation and chemical reactions. He combined copper and sulphur to reduce mercury chloride and reported the combination of sulphur and arsenic with other metals (Howard et al. 2018; Takacs 2018a). His work showed that pressure could be used to form new chemical compounds and presented a good base for further investigations (Parker 1914). On the other hand, Lea was more focused on researching the fundamental of mechanical forces (Johnston and Adams 1913; Parker 1914; Kipp et al. 2005; Baláž 2008). One of his major findings was the observation of the transformation of allotropic silver and the reduction of silver halides (Takacs 2003, 2018b; Boldyrev and Tkáčová, n.d.). During these experiments, Lea noted that the changes of the two compounds did not depend on heat. When applying pressure, a developable latent image was produced in silver halides, while in allotropic silver there was a visible transformation. Consequently, he set the ground on the difference between thermochemical reactions and mechanical grinding reactions (Kipp et al. 2005). Lea highlighted that chemical changes can be induced by mechanical actions, while the heat-generated reactions are completely different routes. These observations were proven with further studies focused on the changes observed on silver chloride and mercuric chloride. Silver chloride was found to decompose, while mercuric chloride sublimed under the application of heat onto the two systems (Takacs 2013, 2018b; El-Eskandarany 2020). Oppositely, the two compounds decomposed under the application of mechanical forces. Although the analyses were done under visual observations, Lea pioneered what is now the base of the mechanochemistry mechanism. Some scientists even called him "the father of mechanochemistry" (Kipp et al. 2005; Parker 1914).

Most of the explorations done in these times were all inspired by geology and the understanding of metals (Takacs 2013). Altogether, they focused on the understanding of the impact of pressure and shear over different chemical compounds. In this same motivation, Lea, Spring, and other researchers found that the changes induced by mechanical work can be the base of new or improved technologies. Particularly, fine grinding found an important use at the end of the nineteenth century in the processing of cement clinker, ores, and powder metallurgy. A prominent researcher called Flavitsky was so passionate by the chemical reaction of solids by grinding that he even extended his research to 36 solid chemicals (Boldyrev, n.d.). By rubbing these solids, he identified different anions and cations. He then compiled a "pocket laboratory" introducing mechanochemistry for the first time in the teaching laboratory (Takacs 2013).

Later, Ostwald coined the term mechanochemistry when he showed the different modes into which energy could be introduced into a system to induce chemical reaction (Kipp et al. 2005; Šepelák et al. 2013). Benjamin, on the other hand, focused on the formation of alloys and of "mechanical alloying", which was developed to make homogeneous composites particles (Benjamin 1990, 1974, 1970). Heinicke provided the accepted definition of mechanochemistry as "the branch of chemistry that deals with the chemical and physical transformation of solids that are induced by the action of mechanical influences". The first international conference in mechanochemistry was then held in the Slovak Republic in 1993, as shown in Figure 5.3. From then, mechanochemistry has been explored in various areas and recently it was extended to the extraction of specific component in food.

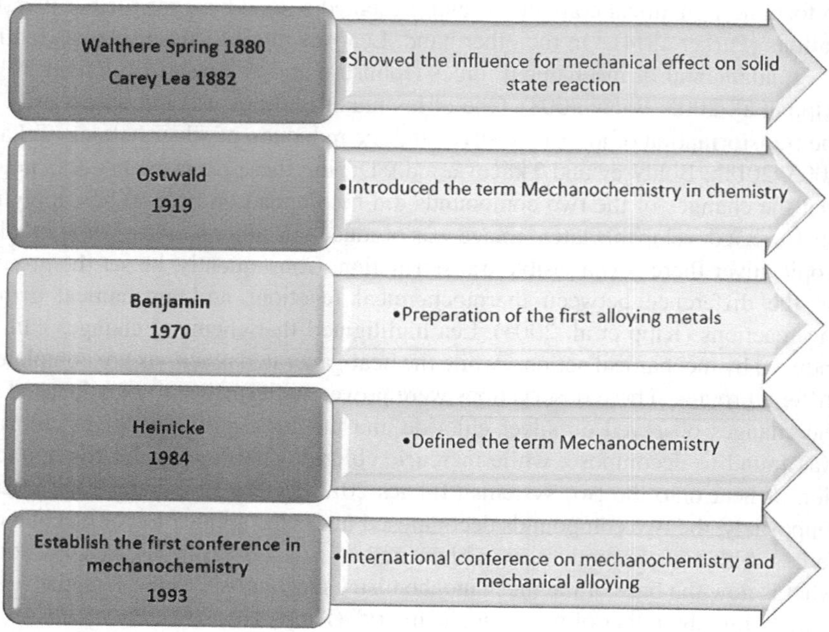

FIGURE 5.3 Historical highlight of the development of mechanochemistry.

Source: Adapted from Kipp et al. (2005).

5.3 EXTRACTION MECHANISM OF MCAE

The process of MCAE can be seen through the deliberate application of mechanical forces through high-energy ball-milling, which led to the preparation or extraction of various substances with excellent characteristics. The direct heating of a sample has different chemical effects compared to mechanical actions obtained in MCAE (Takacs 2013). In general, MCAE consists of a mechanical pre-treatment, mechanochemical treatment, and subsequent extraction procedures (see Figure 5.4) (Wu et al. 2017). The pre-treatment referred to the preparation of the samples which depends on the type of material analysed. The mechanochemical treatment is carried with a solid reagent where the raw material is ground under high-intensive mechanical pressure. The last part is the subsequent extraction which could refer to the removal of a natural oil from the material or the removal of a specific analyte. This can be done by dissolving the material in solvent (can be hot water), which will help in extracting the oil/analyte contained in the raw material (Wu et al. 2017). In this section, some of the mechanisms involve in the MCAE will be described and a broad representation of the process can be observed in Figure 5.4.

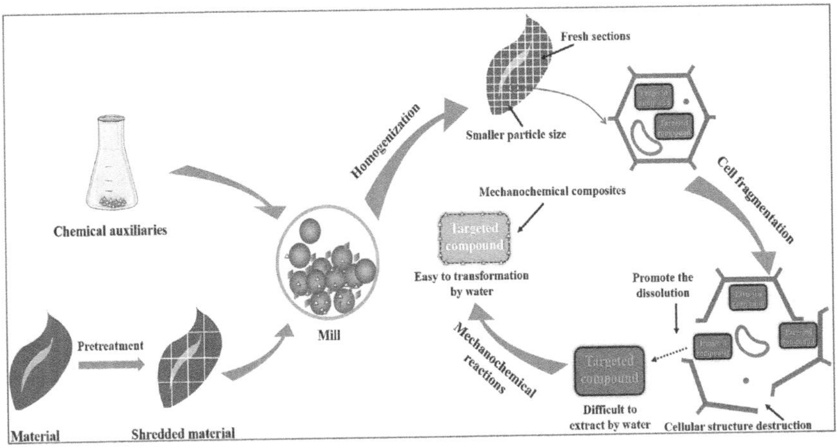

FIGURE 5.4 General concept of MCAE.

Source: Fan et al. (2022).

5.3.1 Homogenization

Particle size is one of the most critical aspects of the mechanism taking place during mechanochemistry. The particle size promotes the access of the chemical reagent to the different components present in the raw material. In MCAE, the particles of the starting material are reduced during grinding in the ball-mill and then homogenized. During this process, there is an increase of the total contact surface area while the particle size decreases. The subsequent process is followed by an improvement of the contact of the particles to the bioactive substances promoting exhaustive and complete reactions (Chemat et al. 2020). Consequently, we have an enhancement of the extraction rate of the targeted analyte.

5.3.2 Cell Fragmentation

One of the hearts of the MCAE lies in the cell fragmentation of raw material to promote the extraction of different analytes. For the extraction of bioactive components in plants, we need to consider that they are protected by a rigid cell wall (Fan et al. 2022). Additionally, when clumped together, the cells are packed, and it is difficult to extract the active ingredients since it will be demanding to release them because of the neighbouring units (Fan et al. 2022). In mechanochemistry, the rigid structure of the cell can be destroyed by the grinding of the starting material. Consequently, cell fragmentation breaks the resistance to dissolution of the bioactive components in the cells while increasing their dissolution rate. The destruction of the cell walls is also accompanied by the disintegration into smaller cellular composites of the organelles (Chemat et al. 2020). The contact between the analyte and the reagent is improved through this process promoting its dissolution rate for further extraction.

5.3.3 Mechanochemical Reactions

Following the fragmentation of the cells, the targeted compounds are released during grinding or are easily susceptible to be extracted (Fan et al. 2022). The next stage is the chemical reactions between the released compounds and the chemical auxiliaries added into the system. The above reaction eases the transfer of the substrate into water facilitating of extraction of the targeted molecules (Lomovsky et al. 2003). In most cases, none or lower quantity of solvent can be used for extraction since the mechanochemical reaction has increased the solubility of the analytes. The following is one of the advantages of MCAE since water can be used to extract medium rendering the experiment less toxic, more cost-efficient compared to other extractions methods (Mateti et al. 2021b). The yield of the experiment is also increased. During the mechanochemical process, the extracted substances are made to react with chemical auxiliaries through an acid and base neutralization process. The auxiliaries have the target to transform the extracted compounds into salts with high solubility (Dobroslavić et al. 2022). Therefore, by adjusting the pH of the system, a selective separation towards specific extracts can be attained. Getting the best output from the experiments carried is important when extracting any substance. To get to that extent, it is necessary to select the most appropriate parameters depending on the analyte to extract.

5.4 MECHANOCHEMICAL METHODS AND INSTRUMENTATIONS

Mechanochemistry is based on the mechanical energy transferred into a specific sample inducing a chemical change within the system. There are different methods of inducing mechanical energy, and depending on the type of sample analysed, a specific method can be chosen.

5.4.1 Mechanochemical Methods

5.4.1.1 Ball-Milling

During the process of mechanochemistry, two stages can be considered (Liu et al. 2022). The first stage is the mass transfer and the formation of heterogeneous reactive contacts. The second stage is the activation of the different contacts by mechanical impact in the ball-mill. During the movement of the grinding balls, which is induced when the milling tank moves, the different mechanical forces are applied directly onto the different materials (Kubota and Ito 2020; Howard et al. 2018). This promotes the breaking of the cell walls in the plants. In general, when the particle size decreases, the molecules store the energy provided by this process. This promotes the chemical reactions which induce the breaking of cell walls of raw materials like plants (Liu et al. 2022).

In ball-milling, mechanochemical effects result from the continuous contact between the different reactants and the balls. The traditional liquid phase processes are quite slow and sometimes the blending is not homogeneous. However, in mechanochemical reactions taking place in a ball-mill, the blending is improved along with filling loading and dispersibility (Hai Nguyen et al. 2019). Furthermore, the process is more successful in obtaining the required material properties. Manual grinding provides a particular force formality which is directly mimicked in ball-milling. Therefore, ball-milling is the first choice for mechanochemistry processes of several chemical compounds like inorganic materials, small organic molecules, and polymers (Liu et al. 2022).

5.4.1.2 Pan-Milling-Induced Mechanochemistry

Most of mechanochemistry equipment are dominated by impact forces which makes it difficult to process some materials. Materials like polymer need to be obtained as superfine compounds; therefore, a more adapted mechanochemical process is needed (Yang et al. 2020). In this case, the use of shear forces is more likely to be applied, and it was proven by Liu et al. (2019), when they studied the preparation of some polymers. A pan-shaped milling consists of a 3D scissors built in a pan-milling. This provided shear forces instead of impact forces like in the ball-milling instrument. In this case, the grinding surface will construct many shearing surfaces which follow the mutual movement of two grinding discs. Therefore, materials are strongly squeezed and sheared, leading to very fine particles (Liu et al. 2022). In results, pan-milling is mostly used in the pulverization of polymer materials since ultrafine particles are formed.

According to Liu et al. (2022), the mechanochemical effects can break hydrogen bonds in cellulose and activate the hydroxyl groups which play an important role in the formation of cellulose derivatives (Liu et al. 2019). This characteristic makes the pan-milling-induced mechanochemistry an excellent technique for application in any processes where very thin particles are needed. Pan-milling-induced mechanochemistry also showed promising application in continuous processing procedures. In this case, it could also be used for scalable application of different compounds through mechanochemistry methodology. In addition, since the pan-milling-induced mechanochemistry produces ultrafine substances due to the mixing involve, the method allowed waste to be recycled. Compared to ball-milling, shearing effect is dependent on the size of the shearing space which limits its efficiency in its application for other compounds like nanomaterials (Liu et al. 2022).

5.4.1.3 Ultrasonication-Induced Mechanochemistry

Ultrasonication-induced mechanochemistry focuses on the study of the compression and expansion of a liquid. Energy is then provided to the liquid phase, inducing the cleavage of chemical bonds leading to chemical reactions. The whole process is based on acoustic cavitation which consists of three different phases. The first phase is characterized by the formation of cavitation nuclei by microbubbles in the liquid. The second phase consist of the growth of microbubbles while the last phase is characterized by the overexpansion of the bubbles formed. The later then leads the bubbles to implode and collapse.

The most advantageous aspect of sonochemical processes is the increase of reaction rates caused by the mechanical effects. Then, mass transfer is accelerated, depassivation and different chemical effects is promoted. In comparison with ball-milling and pan-milling, ultrasonic mechanochemistry is suitable for liquid samples. In ultrasonic mechanochemistry, many chemical reactions are favoured like advanced oxidation processes, induced by the formation of reactive oxygen and radicals. The chemical reactions promoted by this method are polymerization, redox reactions, and degradations, and they are induced by the generation of free radicals through the cleavage of chemical bonds.

5.4.1.4 Instrumentation

A mortar and pestle is used in many households for grinding food to small particles. This applies manual grinding, and it is the simplest instrument employed in most mechanochemical reactions. It is suitable for most laboratory-level reactions due to its need of low-dosage samples and convenience in operation (Liu et al. 2022). Additionally, a mortar and pestle can be used for small trials or tests before proceeding with large-scale experiments. However, this technique is quite exhausting since it necessitates the continuous energy of an individual. The use of a mortar and pestle can also be time-consuming. The aim of the MCAE is to provide an environment where the constant movement results in friction and shear and/or constant collision is maintained. Therefore, it is a necessary to use faster and easier routes to manage mechanical grinding (Fan et al. 2022). Due to the variety of mills found, we can categorize them for laboratory and industrial settings.

5.4.1.5 Planetary Mill

Known for being fast and used for small-quantity grinding, planetary jar is one of the simplest and most used grinding instruments (Abdellaoui and Gaffet 1995; Mio et al. 2004; de Camargo et al. 2019). A planetary jar consists of installed pots on a disc. The disc and the pots are simultaneously rotated in opposite direction at very high speed (Mio et al. 2002). Altogether, the balls in the jar move strongly which then results in the formation of fine grinding of a material derived from a high impact of energy. The grinding is dominated by the transfer of energy from the beads to the materials to be ground (Nilssen and Kleiv 2020). This energy transfer mechanism is called a compressive stressing. According to de Camargo et al. (2019), a planetary mill can be taken as a centrifuge machine where pots containing samples and grinding material are ground while rotating around its own axis and around the axis of the base. Planetary ball-mills are easy to use and versatile; it also has the ability to grind dry or wet materials which expands its applications in various areas (Fan et al. 2022).

5.4.1.6 Vibration Mill

This type of instrument uses external mechanical vibration for the reduction of samples to particles. Mainly, the instrument is mounted with a platform and grinding ball. The platform is usually excited with a frequency and amplitude that could be varied independently (Kiperman et al. 1957). The ball is constantly throwing motion in the milling jar containing the different materials to be ground (Fan et al. 2022). The material collides and is squeezed in the media alternating breakage, compression and split (Fan et al. 2022). One of the most interesting aspects of this instrument is that larger quantity can be ground. Therefore, they found applications in industrial setting due to higher production with low cost of investment. Moreover, the compounds are still efficiently reduced to uniform fine particles with lower consumption (Vedaraman and Raghavendra 1971).

5.4.1.7 Homogenizer

This technique is mostly used for biological materials and worked on the disruption of compound by the use of shear, grind, beat and shock exerted on the samples. A homogenizer is a high-speed system with a small-and-high efficiency (Fan et al. 2022). For analyte like RNA and DNA, this instrument is used in the starting stage of the analysis for the isolation of analytes. Although this type of instrument is unique and can only be used with solvent, the complete release of intracellular components is achieved by a successive beating and multidirectional movement of the ceramic sphere and the sample (Fan et al. 2022). However, they are several parameters that need to be considered and chosen depending on the samples and the analytes to be investigated.

5.5 PARAMETERS OF MCAE

One of the centre areas of MCAE is the parameters used for the extraction of a particular substance. Unfortunately, no specific parameters are attached to the analysis of every sample. However, it all depends on the quantity of a specific analyte in the raw materials. Additionally, the type of method and instrument used will require an adjustment of the milling time, sample preparation, as well as the loading of the sample in the ball-mill. On the other hand, depending on the type of samples, there will be a need of a particular reagent as well as temperature and extraction time. More of these parameters that need to be adjusted are seen in Figure 5.5. In some cases, mismatching parameters may result in the loss of analytes and in low extraction yield. This is caused by the agglomeration of samples which will reduce the accessibility of the mechanical forces to the sample surface area of the particles (Wang et al. 2018).

5.5.1 Sample Preparation

Prior chemical knowledge of the raw sample is required to improve the process of dissociation, dissolution, transformation and transfer of target compounds into

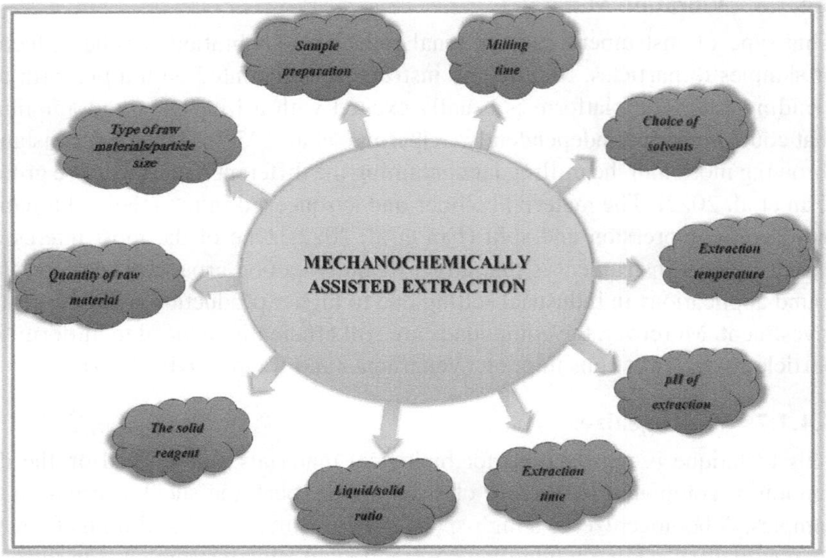

FIGURE 5.5 Different parameters adjusted in mechanochemically assisted extraction.

the extractives. MCAE shows the unique superiority in tissue and cell fragmentation, homogenization of materials. MCAE also shows that the mechanochemical reaction of the dissociated molecules with the auxiliaries, and thus resulting in high yield, at low-cost implication and short extraction time. The MCAE aims to obtain reduced particle size, destroy the cell wall, decompose cellulose, and accelerate the dissolution kinetics that can increase the extraction efficiency and decrease the processing time. Hence, depending on the samples, specific treatment should be done to obtain the best yield (Shahzad Aslam et al. 2017). Therefore, a sample might need to be oven-dried, sliced, crushed, while other samples might just need to be defatted or pre-treated with some other chemicals (Wang et al. 2016; Zhu et al. 2012). These steps will be adapted according to the sample to be processed and the compound that needs to be extracted. In some instant, the analyte would be very unstable, even volatile, and it can be lost during sample preparation. Therefore, a previous study on the analyte and the samples are recommended to obtain the best extraction results (Hernández and Bolm 2017).

5.5.2 Type of Raw Materials and Their Particle Size

As seen in the previous section on instrumentation, the type of materials and their particle size are important parameters. Altogether, the type of raw materials will direct to a specific type of instrument, ball and mill to use. For biological extraction, you will need a homogenizer, while with food raw material a planetary instrument will be preferred. This will have a direct consequence to the size of the particles which is the main part that promotes the breakage of the cells and the easy extraction of the analyte. In general, the smaller the particle inserted in the ball and mill, the lesser time it will require to be ground. The bigger the size of the particles, the more time you will need. Thereafter, the sample preparation which will guarantee the extraction yield. According to Fan and collaborators, fine solid particles are more likely to undergo a better extraction since they could be in full contact with not only the solvent/reagent but also the beads in the grinding vessel. Therefore, pulverized samples can help reduce the cost of an experiment by reducing the milling time. Additionally, the extraction yield and a complete mechanochemical reaction is improved along with the equipment wear (Fan et al. 2022).

5.5.3 Quantity of Raw Material in the Mill

Each mill will be large enough to contain a certain amount of sample. The ratio of the mill to the sample quantity is to be adjusted from the beginning of the experiment. When the sample comes in contact with the grinding balls, it will be reduced if the correct ratio is reached. However, if there is a larger quantity of material in the mill, the impact of the grinding mill will be reduced. This may lead to a lesser yield because the quantity of the material reduces the movement and the collision space as well as the conversion rate (Fan et al. 2022). Therefore, the grinding balls movement is decreased when there is a large amount of material in the milling jar. This will result in a decrease in the extraction rate (Fan et al. 2022). More experimental time might be needed to improve the extraction process of the analyte. On the other hand, when there is less material in the mill, the sample will not be well ground, reducing the fragmentation of the cells, which then leads to the reduction of the analyte released.

5.5.4 The Solid Reagent

During the milling of a material for the extraction of different analytes, a reagent is used to react and ease the breakage of the cell walls present in the component of the raw materials. This is where the chemistry aspect of the mechanochemistry takes place. In this case, the concentration and the type of solid reagent play an important role in MCAE.

The appropriate amount of reagent as well as the type used may increase the extraction efficiency and reduce the waste (Wu et al. 2017). When the amount of the reagent is too low, the process time is increased so the analyte could be neutralized (Korolev et al. 2003; P. Wang and Tang 2009). The kinetic of the mechanochemical process and the yield of the analyte is directly attached to the reaction of the raw material and the reagents. Similarly, when the quantity of the reagent is too high, the analyte is affected. Several research showed that the yield can decline with excess of reagent in the sample (Liu et al. 2007).

When it comes to the choice of reagent, it will depend on the raw material, which can be acidic or basic. For the extraction of acids, the reagents used are solid alkalis. The solid alkali selected will completely depend on its strength and on the analyte to extract. For material with low-acid content, a weak alkali ($NaHCO_3$) will not help in retrieving all the analyte from the sample. The best choice is to select stronger alkali, like NaOH, that will increase the efficiency of the extraction by reacting with all the low-acidic analyte present in the material (Xie et al. 2013). The opposite will happen for material with higher quantity of acid. However, some reagents have been observed to react with the sample leading to undesired products (Wu et al. 2017). Therefore, it is important to study the material to be analysed, which will then help in the choice of relevant reagents and it will likely determine the kinetics of mechanochemical process (Liu et al. 2007; Xie et al. 2013).

5.5.5 Milling Time

The time needed for the raw material to be ground is called milling time. Milling time is the time spent in the vessel and represents the exposure that the raw materials will have with the reagents and the milling balls to promote the cell fragmentation through mechanical energy. This will be important for the release of the analyte and its yield during the process. Milling time will rely mostly on the type of instrument used and the type of reagents, quantity of raw material, and the analyte to extract from the sample. In terms of instrument, a high-speed grinding instrument might not need a lot of time for the reaction to be successful. It will even be better if the correct and precise quantity of reagent is used. When a raw material is ground longer than it should, the analytes might decompose, and the yield will be low during the extraction phase.

5.5.6 Extraction Time

Extraction time is recorded after milling of the sample with the reagent. The extraction time corresponds to the time taken for the analyte to be removed from the ground material after leaving it in a solvent. In general, the ground material is mixed with a solvent with the aim of removing the targeted analytes, and the final solution is centrifuged to separate the different components. The extraction time has an effect on the resources provided. The shorter the extraction time, the less efficient the method employed. However, the longer the extraction time, the more efficient the method and the higher the yield of the analyte. The extraction of any analyte is dependent on the diffusion rate and the dissolution process of the compounds in the solvent. The main advantage of mechanochemistry is the creation of a more soluble compound which will easily dissolve in solvent and enhance the extraction of the analyte.

On the other hand, a longer extraction time can also lead to the decomposition of some analyte, thereby decreasing its yield. There are substrates that are very unstable in solution and might only have a very short life span. Therefore, when left for a much longer time in the solvent, it might lead to different compound or be lost by decomposition. Choosing the correct extraction time will mostly be based on the quantity of the analyte present in the solvent after extraction and it could impact the efficiency of other parameters (Fan et al. 2022).

5.5.7 Extraction Temperature

Extraction temperature is usually targeted with the extraction type since the temperature of a system can affect the dissolution and diffusion rate of the constituents in the solvent. The higher the temperature, the higher the dissolution and diffusion rate of the analyte in the solvent. At high temperature, kinetic of the particles is increased leading to faster mixing time in the solvent. This can be important for compounds that are difficult to dissolve in water or other solvents at room temperature. By increasing the temperature, the extraction yield will be increased by improving the dissolution of the analyte in solvent. One of this application could be observed in traditional preparation of palm oil. In some culinary process, hot water is used as a solvent to extract palm oil from the ground palm oil nuts. Then oil is extracted and used for further food preparation.

5.5.8 pH of Extraction

The pH of the solution will have an direct impact on the yield of the analyte. Depending on the type of material you are extracting (acid, base, or neutral), it might be needed to adjust the pH before extraction. In case of the use of an alkali reagent, after grinding the raw material and dissolving it in a solvent, the solution may be more basic. For analytes that are acidic, the yield will be affected due to the alkali medium. Once obtained, the solution needs to be adjusted to an acidic medium to improve the yield of extraction for the acidic analyte. In MCAE, the targeted compound is a salt, therefore for subsequent processes of the acid and base reaction, adjusting the pH will help remove extracts while improving the transformation of the analyte. In some other cases, adjusting the pH can also help in removing impurities in the sample.

5.5.9 Choice of Solvents

Following the mechanochemical treatment of the materials, the chemical reaction produces several water-soluble salts, complexes, and other hydrophilic compounds. As a result, in most situations, water is chosen as the primary extraction solvent. In addition, due to the low polarity of several target molecules, ethanol was used as an extraction solvent in some studies. For a cost-efficient process, water is the most appealing solvent for analyte extraction. The fact that salts are formed during the mechanochemical process helps in the use of water as solvent for extraction.

5.5.10 Liquid/Solid Ratio

Generally, as the liquid/solid ratio increases, the extraction efficiency of chemicals improves. A high liquid/solid ratio may require more time in the subsequent concentration procedure, and the high temperature created during the concentration process may have an effect on the active components. Using the yield of the targeted extract, some researchers determined the appropriate liquid/solid ratio.

In summary, several parameters need to be adjusted to improve the extraction of compounds. Every parameter will have an effect towards the final yield of a particular substance during the extraction. As seen in Table 5.1, the frequency of collisions of the reactants is directly affected by the number of balls used as well as particles.

5.6 PRACTICAL APPLICATION OF MCAE IN ESSENTIAL OIL EXTRACTION/DERIVATIVES AND OTHER COMPOUNDS

As the need for green technologies increases, MCAE is shown to be the potential substitute over other methods for extraction in Table 5.1. The low cost and high yield of this method have been associated with the increase of its application around the world. Additionally, some investigations have also highlighted its mild experimental condition and short extraction time, and the ease to integrate the method in subsequent analyses has promoted its practice (Fan et al. 2022).

MCAE has been applied for the extraction of many bioactive compounds and essential oils. In various African countries, grinding is applied for the generation of palm oil in different households. In general, the palm oil nuts are washed, ground with a mortar and pestle until all the fibres are separated from the nucleus of the nuts. Following that, the oil is extracted using boiled water as a solvent and then used for any cooking preferences. This is a direct application of MCAE. After the first application of MCAE for triterpenoid acids extraction from the Siberian fir needles, several analyses have been done to prove its usefulness in different MCAE (Fan et al. 2022).

Goncharov et al. (2006) have improved the extraction of lappaconitine from aconitum septentrional roots. Prior to the chemical treatment, the roots were dried and then finely ground in a planetary mill. They eventually analysed the effect of isopropanol, chloromethane, ethanol and a mixture of chloromethane and gamma alumina for extraction. The highest extraction was obtained with the combination of the solvent/sorbent with an extract of 63%. The amount of lappaconitine was found to double in this process. The lappaconitine extracted was quantified using a column chromatography.

Lomovsky, Korolyov, and Kwon (n.d.) used MCAE to extract several bioactive ingredients contained in different plants. They found that by including the mechanochemical approach, they successfully extracted triterpenoid acids, phytosterols, and ecdysteroids. In many cases the yield of the bioactive ingredient was increased by more than 50% and its purification was ameliorated. The following analysis renders the MCAE an efficient method for the selective removal of green products and to improve their purification.

Isofraxidin was extracted from *Eleutherococcus senticosus* (Siberian ginseng), a plant used in some countries to stimulate the immune system. The plant was chopped and blended then ground for some time. The most interesting aspect of Liu et al.'s investigation is the study of different parameters to reach the best extraction. The finding showed that the MCAE was more efficient than the heat-reflux method. Indeed, compared to conventional methods, the MCAE gave

TABLE 5.1
Different application of MCAE as a extraction technique.

Sample type	Extracted compound	Operating conditions	Type of milling	References
Fur needles	Triterpene acids	Centrifugal activator: 200 m/s^2 Steel ratio: 170:5 Treatment time: 3 min	–	Korolev et al. (2003)
Ginkgo leaves	Flavonoids Terpenetrilactones	Solid reagent ($NaHCO_3$) 21%, milling time 7.5 min, and solvent/solid ratio of 33 ml/g	–	Zhu et al. (2011)
Citrus aurantium L. var. *amara* Engl.	Essential oils	5% fumed silica A200 as solid reagent, ball-milling at 300 r/min for 10 min, and extraction with *n*-hexane (1 g/5 ml) during 30 min	Planetary ball-mill	Zhu et al. (2022)
Bamboo (*Phyllostachys edulis*) leaves	Flavonoids	Na_2CO_3 (12wt.%) and $Na_2B_4O_7{\cdot}10H_2O$ (2wt.%) for 10 min, extraction with water at 25°C, and liquid/solid ratio of 20ml/g for 15 min	Planetary ball-mill	Zhu et al. (2011)
Ganoderma lucidum	Polysaccharides	Material/solid reagent (Na_2CO_3)5gÆg)1, milling time 20 min, solution/material ratio of 20 (m/g)1), and extraction time 130 min	Planetary ball-mill	Yang et al. (2017)
Eleutherococcus senticosus	Isofraxidin	Na_2CO_3 content of 2.0% (w/w) and extraction with water (liquid/solid ratio of 20:1, ml/g) for 5 min at 25°C	Vibrational mill	Liu et al. (2007)
Camellia oleifera Abel. meal	Kaempferol Glycosides	NaOH of 20:1 (g/g), a milling time of 15 min, and a solution/solid ratio of 20:1 (ml/g) for 60 min	Planetary ball-mill	Zhu et al. (2011)
Hibiscus mutabilis L.	Rutin	Na_2CO_3 (15.0 wt.%) and $Na_2B_4O_7{\cdot}10\ H_2O$ (1.5 wt.%) for 4 min, extraction temperature at 25°C, total extraction time of 15 min (two cycles of 10 min and 5 min), acidification pH 5.0, and solvent/material ratio of 25 ml/g	Planetary ball-mill	Xie et al. (2011b)

(Continued)

TABLE 5.1 (*Continued*)
Different application of MCAE as a extraction technique.

Sample type	Extracted compound	Operating conditions	Type of milling	References
Magnolia officinalis	Magnolol	Milling with Na_2CO_3 (2.0 wt%) for 7 min, extraction by water at 40°C, and liquid/solid ratio of 25:1 ml/g for 20 min, and precipitation at pH 3.5	Planetary ball-mill	Xie et al. (2011a)
Stephania tetrandra S. Moore	Angchinoline and tetrandrine	Obtained within 15 min of the total extraction time under the optimal conditions: oxalic acid (20 wt %), grinding plant materials for 5 min, and the liquid/solid ratio of 20 ml/g	Planetary ball-mill	Wang et al. (2018)
Watermelon seeds	Protein	Odium bis (2-ethylhexyl) sulfosuccinate (AOT) concentration of 0.10 mol/L, water content of 20, ionic strength of 0.05 mol/L, aqueous phase pH of 7, solid/liquid ratio of 0.2 g/ml, ball-milling speed of 523 rpm, and ball-milling time of 10.5 min	Planetary ball-mill	Liu and Xi (2021)
Sea buckthorn pomaces	Flavonoids	Na_2CO_3 40%, ball-to-material rate 29:1 g/g, milling speed 410 rpm, milling time 24 min, extraction temperature 25°C, extraction time 20 min, and the solid/solution ratio of 1:10 g/ml	Planetary ball-mill	Guo et al. (2021)
Petals of *Crocus sativus* L.	Polysaccharides	Milling for 8.14 min, keeping in ultrasonic bath at 408°C with NaOH (0.2 wt %) for 15 min, and the solid/liquid ratio of 1/30 g/ml	Planetary ball-mill	Yang et al. (2017)

the highest isofraxin yield with a much lower extraction time and energy costs. The same result was obtained in 2012 by Zhu and colleagues who investigated the extractions of polysaccharides from glucidums pores. The extraction was successfully done with the polysaccharide acquired, demonstrating significant antioxidant activities.

Recently, Zhu et al. extracted for the first time an essential oil from *Citrus aurantium* L. var. *amara* Engl. (Rutaceae). A plant grown in Asia which possesses antioxidant, anti-inflammatory, lipid-lowering, anti-atherosclerosis, and anti-tumor pharmacological properties. The oil was successfully extracted with 26 times higher yield than other methods. They proved that the MCAE is an efficient method for the removal of essential oil from *Citrus aurantium* L. var. *amara* Engl. by grinding and exposing the material to solvent extraction.

Chondroitin sulphate is a basic block which formed cartilage in joints. It can be found in humans and animals. It is usually consumed to help with osteoarthritis, a most common type of arthritis. Wang and Tang successfully extracted chondroitin sulphate from shark cartilage using mechanical pre-treatment and mechanochemical treatment (P. Wang and Tang 2009). The full processes were successfully attained with high-intensity planetary activator at room temperature under solvent-free condition. However, when comparing the two methods, Wand and Tang found that the mechanochemical treatment had a higher purity of product with 95.1% of extracted chondroitin sulphate, while the mechanical pre-treatment only resulted in 86.4%. Additionally, the yield was increased by 9.33% with the mechanochemical treatment. In this case, the extraction time was decreased from 3 hours to 3 minutes, which shows the high efficiency of the technique.

Jin et al. extracted Eleutheroside B from *Eleutherococcus senticosus* by mechanochemical pre-treatment (Jin et al. 2012). Eleutheroside B accounts for 45% of acanthopanax extracts. These extracts are known to improve the sleep, reduce fatigue and elevate the efficiency of brain functioning. In their study in the extraction of Eleutheroside B, Jin and collaborators successfully used MCAE. For this research, they chose hydroxypropyl-β-cyclodextrin (HP-β-CD) with sample particle size of less than 44 um (attained after screening parameters). As solvent for extraction, they used water with a temperature of 40°C. The later parameters were found to lead to a short extraction time and a simple process for the extraction of Eleutheroside B compared to heat-reflux extraction (traditional method). They concluded that the MCAE was a less expensive and eco-friendly technique since water was used as a solvent of extraction, which is not the case for heat-reflux extraction.

5.7 ADVANTAGES AND DISADVANTAGES OF MCAE AND FUTURE OUTLOOK

Chemists are encouraged to find the most proficient and environmentally friendly techniques due to the growing concern of the environmental changes (Egorov et al. 2020b). In this aspect, green chemistry was born and has been a field where most chemists attempt to find greener techniques in science. Anastas and Warner

TABLE 5.2
The principle of green chemistry.

Number	Principle	Short description
1	Prevention	Prevent the generation of waste
2	Atomic economy	Minimize solvents generation during synthesis
3	Safer chemical synthesis	Use low or safer reagents for synthesis
4	Safer chemicals design	Designed chemicals should present the lowest toxicity
5	Use of safer solvents and auxiliaries	Avoid use of solvents and reagents
6	Energy efficiency	None or less energy usage
7	Use of renewable raw materials	Renewable raw materials should be used whenever it is economical
8	Reduction of derivatives	Derivatization processes should be avoided or minimized
9	Catalysis	Use catalysis instead of stoichiometric reagents
10	Degradation products design	Design degradable chemicals
11	Real-time analysis for pollution prevention	Monitor analytical method to avoid formation of hazardous substances
12	Accidents prevention	Minimize potential accidents by choosing the right substances and the process to handle them

Source: Aparecida de Marco et al. (2018) and Anastas and Warner (n.d.).

highlighted 12 principles that need to be considered to improve chemistry techniques, which are given in Table 5.2 (Anastas and Warner n.d.).

Among these twelve principles, MCAE has shown that it could prevent the generation of hazardous waste (Štrukil 2018; Egorov et al. 2020a). Other principles that are the main advantages in MCAE are the reduction of solvents, potential accidents, and the use of toxic reagent (water is mostly used for extraction). By selecting the right parameters, the process of MCAE can be simplified while been extremely efficient. Consequently, MCAE is considered as a green technology, showing distinctive superiority in application of high-energy mechanical force. MCAE acts on the raw materials to dissociate the target components from natural products and improve its solubility in aqueous media under normal temperature. The whole process takes place without the use of any harmful, toxic, and costly organic solvents in most cases.

As interest in bioactive compounds increases, MCAE is regarded as better extraction method with maximum production benefits involving less processing time and lower cost in the ongoing exploration. The main criteria under consideration – maximum intensity and mechanical treatment time – define the technological conditions under which a raw material should be treated to avoid disintegrating the unstable valuable compounds. Previous studies proposed to achieve better yield of up to 50% using mechanochemical treatment. Because it prevents the deterioration of important components, solid-phase treatment is preferable to solvent-based treatments. Many physiologically active compounds, particularly antioxidants, exhibit significant differences in stability between liquid and solid phases. For example, in the presence of a modest amount of liquid phase, the rate of oxidative polymerization is typically 100–1,000 times faster. Solid-phase mechanochemical reactions significantly accelerate degradation processes and improve the production of physiologically active components.

Many organic molecules, particularly those with intricate structures, become unstable when subjected to strong mechanical treatment. Under mechanical treatment, many chemical processes are responsible for the degradation and denaturation of organic molecules. In some situations, molecules are destroyed at covalent bonds by heterolytic or homolytic processes. Heterolytic bond-breaking and radical production result in chain oxidation or organic material condensation; this mechanism is common in organic systems. Homolytic decomposition results in the production of new, lighter chemical molecules. These reactions' by-products reduce the activity of biologically active preparations and may have uncontrollable physiological repercussions. While destroying and reordering hydrogen bonds leaves the molecules in their initial state, a newly generated complex structure may be conserved in solution and modify the biological activity of compound.

Further exploration and application of MCAE is important, including the extraction of other materials, especially water-insoluble substances and heat-sensible compounds. Additionally, further development of extraction equipment is also necessary to improve the efficiency of the method. The ball-mill with different scales available might permit the applicability of the MCAE to a larger scale.

One of the most difficult aspects of MCAE is the versatility of the parameters. The available data are non-consistent, rendering the method not attractive for most users. Additionally, the mechanism and kinetics of MCAE for natural products is still under discussion and needs further investigation. In general, a deeper understanding of mechanism is essential for further development of MCAE and to reduce the screening of specific parameters for the extraction of different substances. For example, the physicochemical and biological changes caused by mechanochemical pre-treatment need to be further investigated. Additionally, the understanding of the reaction kinetics of MCAE requires special attention since it is directly attached to a specific sample. Although MCAE has not yet been applied widely in modern food and pharmaceutical industry, there is a strong belief that it will be a competitive tool in extraction of bioactive substances in the future. The further improvement of the extraction process and equipment will facilitate the application of MCAE technology. Subsequently, competitive tools in extraction of bioactive compounds in the future will require development of temperature-controlled devices for cooling and/or heating.

5.8 CONCLUSION

Mechanochemistry application has been recorded for more than a thousand years ago with the first documented application by Theophrasus (315 BC). Nowadays, mechanochemistry has still been part of many culinary practices in different households around the world. When applied for the extraction of natural compounds, it takes on a much specific path by providing an efficient and cost-effective choice for scientists.

MCAE is a new technology that has gained enormous drive as an alternative procedure to remove essential products from plants or other natural constituents. The increased use of MCAE is mainly focused on the reduction of toxic solvents during the experiments. In other words, the green aspect of the technique and the improvement of the solubility of samples have given more options to scientists to develop. MCAE has the advantage of providing high yield, lower cost, short extraction time, mild experimental condition and ease to connect with other analysis. With almost no need for solvents during mechanochemistry and with a reduced amount of solvent (or use of greener solvent like water) used for extraction, these characteristics have revolutionized this technique in academic and industrial sectors.

However, there are many impact parameters that necessitate adjustment to improve the extraction yield of analytes. Per example reaction time can be affected by the milling time and particle size as well as the type of reagent used during mechanochemistry. Also, the type of mills, the filling quantity, and the reagent can affect the effective mixing of the reactants. On the other hand, the energy transfer can be impacted by the reagent, number of balls, and the type of mills used. Therefore, sample preparation, type of raw materials, type of solid reagent, milling time, and extraction time are a few parameters that play an important role in the extraction of analytes. Determining the right parameters for the analysis of a particular analyte will improve the yield of the experimentation. Although there are many underlying mechanisms that are unknown and need more investigations, MCAE still is an efficient and appealing technique for the extraction of natural products.

REFERENCES

Abdellaoui, M., and E. Gaffet. 1995. The physics of mechanical alloying in a planetary ball mill: mathematical treatment. *Acta Metallurgica et Materialia* 43.

Anastas, Paul, and John Warner. n.d. *Green Chemistry: Theory and Practice.* Oxford University Press. Accessed June 28, 2023. https://global.oup.com/academic/product/green-chemistry-theory-and-practice-9780198506980?cc=gb&lang=en&.

Aparecida de Marco, Bianca, Bárbara Saú Rechelo, Eliane Gandolpho Tótoli, Ana Carolina Kogawa, and Hérida Regina Nunes Salgado. 2018. "Evolution of Green Chemistry and Its Multidimensional Impacts: A Review". https://doi.org/10.1016/j.jsps.2018.07.011.

Baláž, Peter. 2008. *Mechanochemistry in Nanoscience and Minerals Engineering*, Vol. 1. Berlin, Heidelberg: Springer-Verlag.

Benjamin, John S. 1970. *Internal Oxidation Technique 12–17 Involves Exposing*, Vol. 1. New York–London: Plenum Press.

Benjamin, J S. 1974. "Volin are Nickel Alloys Group Research Manager and Research Metallurgist, Metal Physics Section, Respectively, The International Nickel Company, Inc, Paul D. Merica Research Lab-Oratory". *Metallurgical Transactions* 5: 1929–34.

Benjamin, John S. 1990. "Mechanical Alloying – A Perspective". *Metal Powder Report* 45: 122–27. https://doi.org/10.1016/S0026-0657(10)80124–9.

Boldyrev, V. V. n.d. "Mechanochemistry and Mechanical Alloying 2003 Mechanochemical Modification and Synthesis of Drugs". *Journal of Materials Science* 39(16–17): 5117–20

Boldyrev, V. V., and K. Tkáčová. n.d. "Mechanochemistry of Solids: Past, Present, and Prospects". *Journal of Materials Synthesis and Processing* 8.

Bychkov, Aleksey, Ekaterina Podgorbunskikh, Elena Bychkova, and Oleg Lomovsky. 2019. "Current Achievements in the Mechanically Pretreated Conversion of Plant Biomass". *Biotechnology and Bioengineering*. John Wiley and Sons Inc. https://doi.org/10.1002/bit.26925.

Camargo, Italo Leite de, Rogério Erbereli, João Fiore Parreira Lovo, and Carlos Alberto Fortulan. 2019. "Planetary Mill with Friction Wheels Transmission Aided by an Additional Degree of Freedom". *Machines* 7 (2). https://doi.org/10.3390/machines7020033.

Chemat, Farid, Maryline Abert Vian, Anne Sylvie Fabiano-Tixier, Marinela Nutrizio, Anet Režek Jambrak, Paulo E.S. Munekata, Jose M. Lorenzo, Francisco J. Barba, Arianna Binello, and Giancarlo Cravotto. 2020. "A Review of Sustainable and Intensified Techniques for Extraction of Food and Natural Products". *Green Chemistry*. Royal Society of Chemistry. https://doi.org/10.1039/c9gc03878g.

Dobroslavić, Erika, Maja Repajić, Verica Dragović-Uzelac, and Ivona Elez Garofulić. 2022. "Isolation of Laurus Nobilis Leaf Polyphenols: A Review on Current Techniques and Future Perspectives". *Foods*. MDPI. https://doi.org/10.3390/foods11020235.

Egorov, Ilya N., Sougata Santra, Dmitry S. Kopchuk, Igor S. Kovalev, Grigory V. Zyryanov, Adinath Majee, Brindaban C. Ranu, Vladimir L. Rusinov, and Oleg N. Chupakhin. 2020a. "Ball Milling: An Efficient and Green Approach for Asymmetric Organic Syntheses". https://doi.org/10.1039/c9gc03414e.

Egorov, Ilya N., Sougata Santra, Dmitry S. Kopchuk, Igor S Kovalev, Grigory V Zyryanov, Adinath Majee, Brindaban C. Ranu, Vladimir L Rusinov, and Oleg N. Chupakhin. 2020b. "Green Chemistry Critical Review Ball Milling: An Efficient and Green Approach for Asymmetric Organic Syntheses". https://doi.org/10.1039/c9gc03414e.

El-Eskandarany, M. Sherif. 2020. "The History and Necessity of Mechanical Alloying". *Mechanical Alloying*, January, 19–53. https://doi.org/10.1016/B978-0-12-818180-5.00003-0.

Fan, Linhong, Wenxiang Fan, Yuqi Mei, Longchan Liu, Linnan Li, Zhengtao Wang, and Li Yang. 2022. "Mechanochemical Assisted Extraction as a Green Approach in Preparation of Bioactive Components Extraction from Natural Products – A Review". *Trends in Food Science and Technology*. Elsevier Ltd. https://doi.org/10.1016/j.tifs.2022.09.009.

Gligor, Octavia, Andrei Mocan, Cadmiel Moldovan, Marcello Locatelli, Gianina Crişan, and Isabel C.F.R. Ferreira. 2019. "Enzyme-Assisted Extractions of Polyphenols – A Comprehensive Review". *Trends in Food Science and Technology*. Elsevier Ltd. https://doi.org/10.1016/j.tifs.2019.03.029.

Goncharov, A E, A A Politov, N A Pankrushina, and O I Lomovskii. 2006. "isolation of Lappaconitine from Aconitum Septentrionale Roots by Adsorption". *Chemistry of Natural Compounds* 42.

Guo, Zili, Jingya Cheng, Lei Zheng, Wenhao Xu, and Yuanyuan Xie. 2021. "Mechanochemical-Assisted Extraction and Hepatoprotective Activity Research of Flavonoids from Sea Buckthorn (Hippophaë Rhamnoides l.) Pomaces". *Molecules* 26 (24). https://doi.org/10.3390/molecules26247615.

Hai Nguyen, Quoc, Hyeongwoo Kim, Il Tae Kim, Wonchang Choi, and Jaehyun Hur. 2019. "Few-Layer NbSe 2 @graphene Heterostructures as Anodes in Lithium-Ion Half- and Full-Cell Batteries". https://doi.org/10.1016/j.cej.2019.122981.

Hermann, Gary N., Peter Becker, and Carsten Bolm. n.d. "C À HA Midations Very Important Paper Mechanochemical Iridium(III)-Catalyzed C À HB Ond Amidation of Benzamides with Sulfonyl Azides under Solvent-Free Conditions in AB All Mill". Accessed July 3, 2023. https://doi.org/10.1002/ange.201511689.

Hernández, José G., and Carsten Bolm. 2017. "Altering Product Selectivity by Mechanochemistry". *The Journal of Organic Chemistry* 82: 4007–19. https://doi.org/10.1021/acs.joc.6b02887.

Howard, Joseph L., Qun Cao, and Duncan L. Browne. 2018. "Mechanochemistry as an Emerging Tool for Molecular Synthesis: What Can It Offer?" *Chemical Science*. Royal Society of Chemistry. https://doi.org/10.1039/c7sc05371a.

Jin, L., H. Li, F. Wu, X. Li, H. Yu, N. Cui, J. You, Z. Cao, X. Sun, J. Zhang, X. Wang, C. Song, S. Li, and Y. Xu. 2012. "Application of Mechanochemical Pretreatment Prior to Aqueous Extraction of Eleutheroside B from *Eleutherococcus senticosus*". *Industrial & Engineering Chemistry Research* 51 (36): 10695–701.

Johnston, J., and L. H. Adams. 1913. "Effect of High Pressures on the Physical and Chemical Behavior of Solids". *American Journal of Science* 4–35 (207): 205–53. https://doi.org/10.2475/AJS.S4-35.207.205.

Kiperman, S. L., A. A. Balandln, and I. R. Davydova. 1957. "Brief Communications Effect of Fine Vibrational Milling on the Activity of Raney Nickel". *Russian Chemical Bulletion* 6: 1502–4.

Kipp, Stephan, Vladimir Šepelák, and Klaus Dieter Becker. 2005. "Mechanochemie: Chemie Mit Dem Hammer". *Chemie in Unserer Zeit* 39 (6): 384–92. https://doi.org/10.1002/CIUZ.200500355.

Korolev, K G, O I Lomovskii, O A Rozhanskaya, and V G Vasil'ev. 2003. "Mechanochemical Preparation of Water-Soluble Forms of Triterpene Acids". *Chemistry of Natural Compounds* 39.

Kubota, Koji, and Hajime Ito. 2020. "Mechanochemical Cross-Coupling Reactions". *Trends in Chemistry*. Cell Press. https://doi.org/10.1016/j.trechm.2020.09.006.

Liu, Bohan, Yijun Li, Qi Wang, and Shibing Bai. 2019. "Green Fabrication of Leather Solid Waste/Thermoplastic Polyurethanes Composite: Physically de-Bundling Effect of Solid-State Shear Milling on Collagen Bundles". *Composites Science and Technology* 181 (September): 107674. https://doi.org/10.1016/J.COMPSCITECH.2019.06.001.

Liu, Lei, and Jun Xi. 2021. "Mechanochemical-Assisted Extraction of Protein from Watermelon Seeds with Surfactant". *LWT-Food Science and Technology* 142: 111025. https://doi.org/10.1016/j.lwt.2021.111025.

Liu, Xingang, Yijun Li, Li Zeng, Xi Li, Ning Chen, Shibing Bai, Hanna He, et al. 2022. "A Review on Mechanochemistry: Approaching Advanced Energy Materials with Greener Force". https://doi.org/10.1002/adma.202108327.

Liu, Ying, Li Ji Jin, Xiao Yu Li, and Yong Ping Xu. 2007. "Application of Mechanochemical Pretreatment to Aqueous Extraction of Isofraxidin from Eleutherococcus

Senticosus". *Industrial and Engineering Chemistry Research* 46 (20): 6584–89. https://doi.org/10.1021/IE070346J.

Lomovsky, Oleg, Kirill Korolyov, and Young Soon Kwon. 2003. "Mechanochemical Solubilization and Mechanochemically Assisted Extraction of Plant Bioactive Substances". In *Proceedings of KORUS 2003: The 7th Korea–Russia International Symposium on Science and Technology* (Vol. 1), University of Ulsan, Ulsan, Republic of Korea.

Mateti, Srikanth, Motilal Mathesh, Zhen Liu, Tao Tao, Thrinathreddy Ramireddy, Alexey M Glushenkov, Wenrong Yang, and Ying Ian Chen. 2021a. "Mechanochemistry: A Force in Disguise and Conditional Effects towards Chemical Reactions". *This Journal is Cite This*: *Chemical Communication* 57: 1080. https://doi.org/10.1039/d0cc06581a.

Mateti, Srikanth, Motilal Mathesh, Zhen Liu, Tao Tao, Thrinathreddy Ramireddy, Alexey M. Glushenkov, Wenrong Yang, and Ying Ian Chen. 2021b. "Mechanochemistry: A Force in Disguise and Conditional Effects towards Chemical Reactions". *Chemical Communications* 57 (9): 1080–92. https://doi.org/10.1039/d0cc06581a.

Mio, Hiroshi, Junya Kano, and Fumio Saito. 2004. "Scale-up Method of Planetary Ball Mill". *Chemical Engineering Science* 59 (24): 5909–16. https://doi.org/10.1016/j.ces.2004.07.020.

Mio, Hiroshi, Junya Kano, Fumio Saito, and Kantaro Kaneko. 2002. "Effects of Rotational Direction and Rotation-to-Revolution Speed Ratio in Planetary Ball Milling". *Materials Science and Engineering* 332. www.elsevier.com/locate/msea.

Nilssen, Benedicte Eikeland, and Rolf Arne Kleiv. 2020. "*Silicon Powder Properties Produced in a Planetary Ball Mill as a Function of Grinding Time, Grinding Bead Size and Rotational Speed*". Springer. https://doi.org/10.1007/s12633-019-00340-0/Published.

Parker, Leslie Henry. 1914. "CXLII. – Reactions by Trituration". *Journal of the Chemical Society, Transactions* 105 (0): 1504–16. https://doi.org/10.1039/CT9140501504.

Ribas-Arino, Jordi, and Dominik Marx. 2012. "Covalent Mechanochemistry: Theoretical Concepts and Computational Tools with Applications to Molecular Nanomechanics". *Chemical Reviews*. https://doi.org/10.1021/cr200399q.

Šepelák, Vladimir S, Andre Düvel, Martin Wilkening, Klaus-Dieter Becker, and Paul Heitjans. 2013. "Mechanochemical Reactions and Syntheses of Oxides". *Chemical Society Reviews* 42: 7507. https://doi.org/10.1039/c2cs35462d.

Shahzad Aslam, Muhammad, Muhammad Syarhabil Ahmad, M Ayaz Ahmad, Maria Stefanova Atanassova, and Muhammad Ayaz Ahmad. 2017. "Mechanochemical-Assisted Extraction Method on Medicinal Plants: A Brief Review Bio Active Compounds Screening View Project Cyberbullying Behavior View Project AENSI Journals Advances in Environmental Biology Mechanochemical-Assisted Extraction Method on Medicinal Plants: A Brief Review". *Advances in Environmental Biology*. 11. http://www.aensiweb.com/AEB/.

Štrukil, Vjekoslav. 2018. "Synpacts Syn Lett Mechanochemical Organic Synthesis: The Art of Making Chemistry Green" 29: 1281–88. https://doi.org/10.1055/s-0036-1591868.

Takacs, Laszlo. 2000. "Quicksilver from Cinnabar: The First Documented Mechanochemical Reaction?" *The Journal of The Minerals* 52: 12–13.

Takacs, Laszlo. 2003. "M. Carey Lea, the Father of Mechanochemistry". *Bulletin for the History of Chemistry* 43 (1): 26–34.

Takacs, Laszlo. 2012. "Mechanochemistry and the Other Branches of Chemistry: Similarities and Dierences". *Acta Physica Polonica A* 121 (3): 711–14.

Takacs, Laszlo. 2013. "The Historical Development of Mechanochemistry". *Chemical Society Reviews* 42 (18): 7649–59. https://doi.org/10.1039/c2cs35442j.

Takacs, Laszlo. 2018a. "Walthère Spring and His Rivalry with M. Carey Lea". *Bulletin for the History of Chemistry* 28 (1): 14–20.

Takacs, Laszlo. 2018b. "Two Important Periods in the History of Mechanochemistry". *Journal of Materials Science* 53 (19): 13324–30. https://doi.org/10.1007/s10853-018-2198-3.

Vedaraman, R., and N. M. Raghavendra. 1971. "Studies in Vibration Milling*". *Powder Technology* 4 (6): 313–21.

Wang, Man, Wentao Bi, Xiaohua Huang, David Da, and Yong Chen. 2016. "Ball Mill Assisted Rapid Mechanochemical Extraction Method for Natural Products from Plants". *Journal of Chromatography A* 1449: 8–16. https://doi.org/10.1016/j.chroma.2016.04.044.

Wang, Ping, and Junmin Tang. 2009. "Solvent-Free Mechanochemical Extraction of Chondroitin Sulfate from Shark Cartilage". *Chemical Engineering and Processing: Process Intensification* 48 (6): 1187–91. https://doi.org/10.1016/j.cep.2009.04.003.

Wang, Shuling, Rui Zhang, Xiaoyu Song, Mengmeng Wei, Tian Xie, and Jun Cao. 2018. "Mechanochemical-Assisted Extraction of Active Alkaloids from Plant with Solid Acids". https://doi.org/10.1021/acssuschemeng.8b02902.

Wu, Kejiang, Ting Ju, Yong Deng, and Jun Xi. 2017. "Mechanochemical Assisted Extraction: A Novel, Efficient, Eco-Friendly Technology". *Trends in Food Science and Technology*. Elsevier Ltd. https://doi.org/10.1016/j.tifs.2017.06.011.

Xie, Jie, Hongjuan Li, Xingyi Zhu, Ping Wang, and Weike Su. 2011a. "Efficient and Selective Extraction of Magnolol from Magnolia Officinalis by Mechanochemical Extraction Technique". *Chemical Engineering and Processing: Process Intensification* 50 (3): 325–30. https://doi.org/10.1016/j.cep.2011.01.004.

Xie, Jie, Yin Shuang Lin, Xiang Jun Shi, Xing Yi Zhu, Wei Ke Su, and Ping Wang. 2013. "Mechanochemical-Assisted Extraction of Flavonoids from Bamboo (Phyllostachys Edulis) Leaves". *Industrial Crops and Products* 43 (1): 276–82. https://doi.org/10.1016/j.indcrop.2012.07.041.

Xie, Jie, Lixia Shi, Xingyi Zhu, Ping Wang, Yi Zhao, and Weike Su. 2011b. "Mechanochemical-Assisted Efficient Extraction of Rutin from Hibiscus Mutabilis L". *Innovative Food Science and Emerging Technologies* 12 (2): 146–52. https://doi.org/10.1016/j.ifset.2010.12.009.

Yang, Jun, Yingpeng Tong, Kunming Zhu, Yu Jiang, Yongqiu Yan, Suhong, Chen, and Ping Wang. 2017. "Optimization of Mechanochemical-Assisted Extraction and Decoloration by Resins of Polysaccharides from Petals of Crocus Sativus L". *Journal of Food Processing and Preservation*, 1–11. https://doi.org/10.1111/jfpp.13369.

Yang, Shuangqiao, Baojie Wei, and Qi Wang. 2020. "Superior Dispersion Led Excellent Performance of Wood-Plastic Composites via Solid-State Shear Milling Process". *Composites Part B: Engineering* 200 (November): 108347. https://doi.org/10.1016/J.COMPOSITESB.2020.108347.

Zhang, Rui, Yue Wang, Xiaoyu Song, Juan Yang, Xin Dong, Tian Xie, Shuling Wang, and Jun Cao. 2020. "Eco-Friendly Mechanobiological Assisted Extraction of Phenolic Acids and Flavonoids from Chrysanthemum". *Journal of Pharmaceutical and Biomedical Analysis* 186: 113327. https://doi.org/10.1016/j.jpba.2020.113327.

Zhu, Peixi, Mengya Hao, Feng Su, Wenhao Xu, Qihong Zhang, Weike Su, and Erwin Adams. 2022. "Mechanochemical-Assisted Extraction of Essential Oils from Citrus Aurantium L. Var. Amara Engl". *Industrial Crops and Products* 188 (November). https://doi.org/10.1016/j.indcrop.2022.115703.

Zhu, Xing Yi, Hai Min Lin, Xiu Chen, Jie Xie, and Ping Wang. 2011. "Mechanochemical-Assisted Extraction and Antioxidant Activities of Kaempferol Glycosides from Camellia Oleifera Abel. Meal". *Journal of Agricultural and Food Chemistry* 59 (8): 3986–93. https://doi.org/10.1021/jf1042689.

Zhu, Xing Yi, Yi Li Mang, Jie Xie, Ping Wang, and Wei Ke Su. 2011. "Response Surface Optimization of Mechanochemical-Assisted Extraction of Flavonoids and Terpene Trilactones from Ginkgo Leaves". *Industrial Crops and Products* 34 (1): 1041–52. https://doi.org/10.1016/j.indcrop.2011.03.013.

Zhu, Xingyi, Xiu Chen, Jie Xie, Ping Wang, and Weike Su. 2012. "Mechanochemical-Assisted Extraction and Antioxidant Activity of Polysaccharides from Ganoderma Lucidum Spores". *International Journal of Food Science and Technology* 47 (5): 927–32. https://doi.org/10.1111/j.1365-2621.2011.02923.x.

6 Microwave-Assisted Extraction of Essential Oil

Chiazor Faustina Jisieike, Maduka Augustine Nwufo, Abiola Ezekiel Taiwo, Paul Musonge and Ifeyinwa Obiora-Dimson

6.1 INTRODUCTION

Essential oils (EOs) consist of intricate combinations of volatile organic components naturally obtained in plants, and their historical importance is widely acknowledged for their therapeutic, aromatic, and culinary attributes. They find extensive usage in diverse industries, including pharmaceuticals, food processing, cosmetics, perfumery, and natural medicine. Typically, EOs are extracted from diverse plant components, including leaves, fruits, barks, stems, flowers, wood, roots, gums, seeds, rhizomes, heartwood, twigs, and buds (Movaliya, 2017). Each EO possesses a distinct chemical composition that imparts specific qualities. These oils have therapeutic properties, including antimicrobial, anti-inflammatory, antiparasitic, antitumoral, analgesic, and antioxidant effects, making them valuable in natural remedies for common ailments like headaches, muscle pain, respiratory issues, and skin conditions (Liu et al., 2018a; Drinić et al., 2020). EOs readily dissolve in organic solvents, yet they exhibit insolubility in water under ambient temperature conditions (Martínez-Abad et al., 2020). Driven by increased demands for natural and nontoxic substitutes for synthetic chemicals, researchers are actively exploring efficient approaches to extracting valuable compounds from plants, as the process of extracting these oils from plant sources plays a vital role in harnessing their essential constituents. The choice of method influences the EO yield, the preservation of specific volatile compounds, and the concentration of bioactive constituents.

Conventional extraction methods like solvent-based extraction, steam distillation (SD), and cold pressing (CP) often involve prolonged extraction periods, significant solvent consumption, and frequently result in low yields (Calinescu et al., 2017; Liu et al., 2018a; Martínez-Abad et al., 2020; Bagade and Patil, 2021). Furthermore, traditional techniques are not well suited for contemporary industries seeking cost-effective and sustainable extraction approaches, given their time and energy demands. Methods such as supercritical fluid extraction

 DOI: 10.1201/9781032634180-6

and ultrasound (US)-assisted extraction are recognized for their efficiency and selectivity. However, their potential limitations lie in the need for specialized equipment and significant investment, which can influence their suitability for various applications (Liu et al., 2018a). Lately, the growing interest in the adoption of microwave-assisted extraction (MAE) technology for EOs extraction has garnered significant interest as an attractive substitute for traditional extraction approaches. This is primarily due to its capacity for volumetric heating, maximal penetration of electromagnetic energy into biological materials, reduced processing time, low solvent consumption, high extraction efficiency, simplified equipment, controllable heating process, moderate initial investment cost, and excellent performance under ambient conditions (Lee et al., 2016; Movaliya, 2017; Sadeghi et al., 2017; Hou et al., 2019; Parizotto et al., 2019; Bagade and Patil, 2021). This technique has arisen as an environmentally friendly and sustainable solution, presenting possibilities for extensive implementation (Ciriminna et al., 2016; Liu et al., 2018a). Its capacity for volumetric heating and rapid deactivation of the heat source significantly impact both product quality and energy consumption, thereby influencing the economic aspects of production. This technology is utilized in a wide range of settings, spanning from analytical laboratories to industrial extractors.

MAE applies electromagnetic energy to enhance the extraction process through the utilization of dielectric heating principles (Parizotto et al., 2019). When exposed to microwaves (MWs), polar molecules rapidly align themselves in response to the oscillating electric field generated by the MWs. As the field direction changes, the polar molecules consistently strive to reorient, creating molecular friction and generating heat. The generated heat causes the bonds between the target compounds and the sample matrices to break, facilitating the release of the desired substances into the extraction medium. The unique ability of MWs to selectively heat polar components accelerates the extraction process while minimizing thermal degradation. This aspect is particularly important for EOs, which often contain delicate and volatile compounds susceptible to degradation under high temperatures (Franco-Vega et al., 2019). By utilizing this approach, mass transfer and extraction kinetics of the target compounds are improved, leading to significant time and energy savings (Fiorini et al., 2020).

Many research teams have successfully utilized MAE to obtain EOs with enhanced purity and reduced cost, resulting in significant yields (Deepak et al., 2013; Kusuma and Mahfud, 2017; Liu et al., 2018a; Drinić et al., 2020). For instance, Liu et al. (2018a) achieved a high yield of 3.19% when extracting EO from lavender at optimal conditions of 500 W MW power, a liquid-to-solid ratio of 17 ml/g, and 40 minutes extraction time. In contrast to hydro-distillation (HD), which yielded 0.91% (w/w), Deepak et al. (2013) achieved a maximum yield of 1.2% (w/w) of eucalyptus citriodora oil using MAE. Drinić et al. (2021) obtained an EO yield of 1.5% from *Zingiber officinale* Rosc. using the HD method. This yield was lower compared to the 1.85% obtained by the MW-assisted HD (MAHD) method, which also led to an approximately 80% reduction in extraction time.

The utilization of MW extraction techniques has proven effective in mitigating environmental effects by lowering carbon dioxide (CO_2) emissions into the atmosphere (Cardoso-Ugarte et al., 2013; Drinić et al., 2021). A study by Drinić et al. (2020), utilized MAHD to extract EO from *O. vulgare* L. ssp. *hirtum*, revealing several advantages of MAHD in contrast to traditional HD. MAHD significantly reduced extraction times, ranging from 24 to 45 minutes, compared to the 136 minutes required by HD. Furthermore, MAHD yielded higher quantities of EO, ranging from 2.55% to 7.10%, in contrast to the 5.81% obtained through HD. Moreover, MAHD resulted in EO with elevated levels of oxygenated compounds, accounting for 78.89% to 85.15% of the composition, while HD produced oil with 76.82% oxygenated compounds. Notably, MAHD demonstrated a more eco-friendly approach, consuming electrical power within the range of 0.135 to 0.240 kW h, whereas HD consumed 1.360 kW h. The versatility of MAE extends beyond its conventional use for plant-based extractions. It has successfully demonstrated its efficacy in extracting natural pigments and bioactive compounds from medicinal plants, as well as antioxidants and other valuable substances from a diverse range of matrices (Sadeghi et al., 2017; Chutia and Mahanta, 2021).

This chapter aims to provide an in-depth exploration of MAE, specifically for EOs. It will cover the underlying principles of using MW energy, optimization strategies, and factors influencing extraction efficiency in terms of both quality and yields. The chapter will highlight the advantages and challenges associated with this technique and discuss its potential applications in various industries. Additionally, economic and environmental considerations linked to MAE will be evaluated.

6.2 MECHANISM OF MW HEATING

MWs constitute a form of non-ionizing electromagnetic radiation, characterized by a frequency spectrum that spans from 300 MHz to 300 GHz, with wavelengths falling between 0.001 and 1 m (Chaturvedi, 2018; Bagade and Patil, 2021). They are positioned between infrared waves and radio waves on the electromagnetic spectrum. MWs initiate molecular heating through the combined mechanisms of ionic conduction and dipole rotation (Drinić et al., 2020; Bagade and Patil, 2021). In this process, charged ions within the substance move and align with the alternating electric field (ionic conduction), while polar molecules rotate to align with the field (dipole rotation) (Chutia and Mahanta, 2021; Drinić et al., 2021). Both processes contribute to the generation of heat during dielectric heating. The rapid rotation of polar molecules also leads to intermolecular friction and collisions, further enhancing heat generation within the substance. Moreover, the extent of this phenomenon is directly influenced by the dielectric constant of the medium being employed, as it affects how it absorbs and dissipates electromagnetic energy when exposed to it. The relative dielectric constant of plant matrices is largely influenced by moisture contents (Tang and Resurreccion, 2009). The higher the concentration of polar molecules in a substance, the greater the amount of heat generated.

6.2.1 Working Principle of MAE

When plant material is exposed to MWs, the electromagnetic waves interact with the polar molecules within the plant matrix, leading to their excitation. These molecules experience a reorientation in a direction opposite to that of the applied electric field. This realignment occurs as the electric field's polarity changes at a specific frequency. The rotational movement of the polar molecules generates heat in the surrounding area. The heat generated speeds up the extraction process by raising the temperature of the solvent and facilitating the dissolution of EO constituents present in the plant matrices. Furthermore, the alternating electric field prevents the polar molecules from attaining a state of equilibrium throughout the extraction procedure. As a result, the polar molecules store potential energy while they are within the field. Upon deactivation of the MW energy, the polar molecules gradually relax and return to their equilibrium position. During this relaxation phase, the excess kinetic energy accumulated within the molecules is released in the form of heat (Chutia and Mahanta, 2021). This energy interchange leads to the extraction of EOs from plant matter. It is worth noting that MW excitation does not alter the structure of organic molecules, and the interaction between these molecules is purely kinetic in nature. The way a material heats up when subjected to MWs depends on its shape, size, dielectric constant, and the specific characteristics of the MW equipment employed (Bradshaw et al., 1998). MW power applications frequently employ the frequency ranges of 915 MHz on an industrial scale and 2.45 GHz in household ovens (Desai et al., 2010; Bale and Shinde, 2013; Cardoso-Ugarte et al., 2013). These frequencies enable the effective penetration of biomaterials, such as plant cells, enabling interaction with polar molecules and ultimately facilitating the release of the desired compounds from the sample matrices. In general, the utilization of MW energy stimulates molecular excitation and rotational motion, thereby facilitating in extracting valuable EO compounds from plant materials efficiently and in a relatively brief time frame.

6.2.2 Designs of MAE Reactors

In laboratory settings, it is common to find two distinct types of MW ovens categorized by their cavities: single-mode MWs (also referred to as mono-modes, focused modes, or open systems) and multi-modes MWs (closed systems or pressurized systems) (Mason et al., 2011; Delazar et al., 2012).

6.2.2.1 Single-Mode MW Systems

In single-mode configurations, MW cavities are designed to evenly distribute MW energy to a specific location. These cavities generate only a singular frequency, thereby enabling the excitation of just one resonance mode (Mason et al., 2011; Routray and Orsat, 2012). An open setup mainly consists of a focused MW emitter, a solvent-filled extraction vessel, and a water condenser (Delazar et al., 2012). In an open system for MW extraction, the sample and solvent within the extraction vessel are specifically subjected to MW irradiation. Efficiently harnessing energy from the MW ensures fast heating of the medium, expediting the extraction of the target compound. This system functions under standard atmospheric pressure, with the highest achievable temperature influenced by the vaporization threshold of the selected solvent for extraction (Desai et al., 2010; Delazar et al., 2012). Open systems offer heightened safety compared to closed systems since they can operate under atmospheric pressure and permit the introduction of reagents at any stage throughout the treatment procedure (Delazar et al., 2012). Mono-mode systems, while limited in their capacity to extract larger volumes and quantities of samples, are well suited for small-scale extraction and research purposes.

6.2.2.2 Multi-Mode MW Systems

These systems enable the utilization of multiple resonance modes, facilitating the application of various incident waves onto the sample (Routray and Orsat, 2012). A closed system consists of extraction vessels, a rotating platform, an enclosed chamber, reflectors, and an MW source as its integral elements. It can efficiently extract larger amounts of raw material by utilizing different extraction vessel configurations. Extraction within a multi-mode MW oven is conducted with precise control over pressure and temperature conditions (Desai et al., 2010; Delazar et al., 2012). Due to the heightened pressure within the extraction vessels, resulting in a higher evaporation threshold for the extraction solvent, closed systems can achieve higher temperatures more readily (Delazar et al., 2012). A turntable is frequently employed to ensure uniform distribution of incident waves within the oven, and this setup makes scaling up the system more convenient. In a closed system, there is minimal loss of volatile compounds, and a reduced amount of solvent is necessary. Simultaneous extraction of multiple samples can be effortlessly accomplished using a multi-mode MW extraction system. Multi-mode ovens are suitable for larger-scale extractions and industrial applications. The primary drawback of this system is the significant risk connected to employing high pressure (Delazar et al., 2012).

6.3 ADVANTAGES AND LIMITATIONS OF MAE

The benefits of MAE have been extensively demonstrated since its initial inception, and the technology has evolved considerably over the years. MAE significantly reduces the extraction time when compared with traditional techniques like HD, SD, and Soxhlet extraction (Calinescu et al., 2017; Liu et al., 2018a; Martínez-Abad et al., 2020; Bagade and Patil, 2021). This accelerated extraction process is credited to the rapid and uniform heating of the sample, promoting the swift liberation of EO components. MW dielectric heating, which targets the moisture inherent in all natural materials, is notably more precise, efficient, and selective compared to conventional heating methods (Calinescu et al., 2014; Ciriminna et al., 2016; Franco-Vega et al., 2019; Parizotto et al., 2019). The generation of heat within the specimen as a result of the MW field relies on the existence of a dielectric material, and the liberation of heat becomes evident when the sample experiences dielectric or conducting losses upon exposure to MW radiation. Polar solvents like water exhibit substantial dielectric losses, with their permanent dipole moments being significantly influenced by the strong absorption of MWs. On the contrary, nonpolar solvents like hexane remain unresponsive to MW exposure, and they are classified as MW-transparent solvents. Due to its effective generation of volumetric heat, MW energy stands as a superior choice over various thermal processes (Ramanadhan, 2005; Deepak et al., 2013; Drinić et al., 2021). Unlike classical conductive heating, heat is transferred directly to the material at the same time (from inside outward) (Ciriminna et al., 2016). Moreover, the highest level of electromagnetic energy penetration occurs within the MW frequency range (Desai et al., 2010).

Another significant benefit of MWs over traditional thermal methods is their superior controllability (Ramanadhan, 2005). In processing scenarios, the capability to promptly deactivate the heat source or control it through mild increments of heating has a substantial impact on product quality and, consequently, production economics. The controlled heating in MAE can target specific compounds, enabling the selective extraction of desired EO components while minimizing the extraction of unwanted substances (Calinescu et al., 2014). Due to its automation, MAE enables comprehensive control over extraction variables like time, power, and temperature (Desai et al., 2010). This is particularly useful for large-scale production and research purposes. The shorter extraction time and controlled heating in MAE minimize the exposure of EO compounds to high temperatures, helping to preserve thermally sensitive compounds that might otherwise degrade in longer extraction processes. Additionally, it yields EOs of greater value, enriched with increased quantities of oxygenated compounds (Lucchesi et al., 2004; Desai et al., 2010; Uysal et al., 2010; Calinescu et al., 2014; Filly et al., 2014).

Research investigations have demonstrated that MAE yields higher quantities of extracts with improved extract quality when contrasted with conventional extraction methods (Liu et al., 2018a; Drinić et al., 2020; Martínez-Abad et al., 2020). The utilization of MW energy can effectively rupture cell walls, facilitating the release of EO constituents under elevated temperature and pressure conditions. Consequently, this leads to shorter extraction durations, ultimately enhancing the overall efficiency of the extraction procedure (Franco-Vega et al., 2019; Hou et al., 2019). Furthermore, the decreased heat exposure and shorter extraction times in MAE can result in EOs with enhanced aroma, color, and chemical composition (Ciriminna et al., 2016; Parizotto et al., 2019). Moreover, MAE is highly effective in achieving a substantial recovery of bioactive compounds (Abedi et al., 2017). MAE often requires smaller volumes of solvent compared to traditional methods due to its efficient extraction technique (Parizotto et al., 2019; Bagade and Patil, 2021). Numerous MW-assisted solvent-free reactions demonstrating increased extraction yields have been documented in academic publications (Lucchesi et al., 2004; Uysal et al., 2010; Filly et al., 2014; González-Rivera et al., 2016). This can offer benefits from both environmental and economic viewpoints, owing to decreased process-related hazards in comparison to traditional extraction methods. Particularly in the extraction of EOs, MW-assisted techniques are greatly preferred for their compact equipment dimensions and simplified device design

(Cardoso-Ugarte et al., 2013; Calinescu et al., 2014). MWs are widely employed in diverse commercial applications on a large scale and offers significant returns on invested capital (Filly et al., 2014).

MAE consumes less energy in comparison to traditional methods due to focused and efficient sample heating (Liu et al., 2018a; Fiorini et al., 2020). This reduction in energy consumption required to achieve the desired analyte directly contributes to the reduced production costs of this technique with respect to energy and time (Delazar et al., 2012; Franco-Vega et al., 2019). Additionally, the utilization of MWs in extraction processes brings the extra advantage of minimizing environmental footprint through reduced CO_2 emissions into the atmosphere (Cardoso-Ugarte et al., 2013; Movaliya, 2017). For instance, when extracting 1 g of EO from rosemary, MAE utilizes 0.25 kWh of energy, while HD requires 4.5 kWh. Furthermore, the time needed for HD extraction is 4 hours, whereas MAE efficiently accomplishes the process within 30 minutes. Notably, the environmental impact varies significantly between the two methods, with HD emitting 3,600 g CO_2/g oil in contrast to solvent-free MW extraction (SFME)'s significantly lower emission of 200 g CO_2/g oil (Filly et al., 2014). MAE also exhibits remarkable versatility regarding the choice of vessels, sample sizes, temperature, pressure, solvent quantities, and sample numbers (Delazar et al., 2012). This adaptability has positioned MAE as an attractive technique for effectively facilitating high-capacity sample processing and routine extraction of natural compounds from a diverse array of botanical sources (Ciriminna et al., 2016).

Several limitations have been identified in the application of MAE. The use of elevated temperatures could lead to the degradation of heat-sensitive bioactive compounds. Additionally, extra clean-up procedures, such as centrifugation or filtration, are required to eliminate solid remnants post extraction (Movaliya, 2017; Chaturvedi, 2018). MW efficiency might be compromised when dealing with nonpolar compounds or solvents, and this also holds true for volatile substances. Opting for solvents that readily absorb MWs is often the preferable approach. Users of both MW ovens and industrial MW systems often encounter the challenge of nonuniform heating, which can be impacted by several variables, including the design of the MW chamber, discrepancies in material properties, and variations in sample size (Tang and Resurreccion Jr, 2009).

6.4 ADVANCES IN MAE TECHNIQUES

The primary reasons for considering the adoption of MW energy are to expedite the process, enhance quality, minimize cost, and maximize yield (Datta and Rakesh, 2013). In recent times, there has been a growing trend toward the adoption of enhanced extraction methods involving MW heating systems, such as MW steam distillation (MSD), MAHD, SFME, and MW-assisted hydrodiffusion and gravity (MHG). The choice of technique used has a notable impact on the yield of EO, preservation of volatile components, and concentration of bioactive constituents (Drinić et al., 2020). The following techniques have been developed.

6.4.1 MW-Assisted Hydro-Distillation

MAHD is recognized as a unique HD approach that creatively combines MW energy and water to efficiently extract desired components from plant materials, as shown in Figure 6.1. It presents itself as an eco-friendly extraction method as it eliminates the need for typical organic solvents commonly utilized in Soxhlet, US-assisted extractions (UAEs), and accelerated solvent extraction (Kusuma and Mahfud, 2017). The effectiveness of MAHD is greatly influenced by the dielectric constants in both water and the plant material (Golmakani and Moayyedi, 2015). Several authors have documented the primary benefits of MAHD for EOs, which include decreased extraction duration, improved yields, increased concentrations of oxygenated compounds, eco-friendly attributes, energy-savings, and economic viability (Racoti et al., 2017; Liu et al., 2018a; Drinić et al., 2020, 2021). MAHD can be conducted by connecting MW systems (single-mode or multi-mode) to the Clevenger-type apparatus, which maintains continuous condensation of the distillate (Liu et al., 2018a; Drinić et al., 2020). An external infrared sensor can be employed to monitor the temperature. This approach is well suited for rapid and efficient extraction of volatile bioactive constituents from plant material (Bagade and Patil, 2021). The MAHD process achieved significant

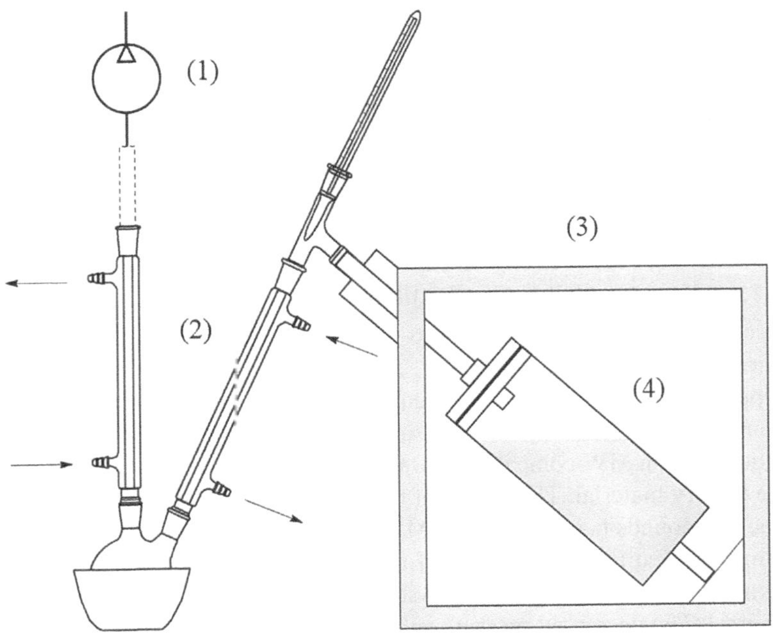

FIGURE 6.1 MAHD apparatus setup: (1) vacuum pump; (2) condensers; (3) 45° rotative Milestone RotoSYNTH MW system; (4) MW vessel.

Source: Copyright obtained from Elsevier (Bustamante et al., 2016).

temperature acceleration by combining MW heating with efficient energy transfer from the electromagnetic field and utilizing an MW-absorbing solvent. As a result, the extraction duration experienced a notable reduction. According to Stashenko et al. (2004), the MAHD method for extracting volatile metabolites from Colombian *Xylopia aromatica* (Lamarck) fruits required only one-fourth the time compared to HD extraction while still achieving the same oil yield. In their study, Liu et al. (2018a) employed a 2.45 GHz multimode MW reactor for MAHD extraction of lavender EOs, optimizing the process with 500 W MW power, a liquid-to-solid ratio of 17 ml/g, and at 40 minutes. The findings showed that MAHD outperformed the traditional HD technique with regard to energy efficiency and duration of extraction, completing the process in just 40 minutes, as opposed to the 120 minutes required by HD. Racoti et al. (2017) employed a TE10n single-mode MW cavity for EO extraction from fresh ginger root using the MAHD technique. Under optimized conditions, they obtained 0.35 g of oil per 100 g of whole ginger root in just 5 minutes with a power consumption of 1,000 W (0.40 kWh/kg). In contrast, conventional HD yielded 0.2 g of oil per 100 g ginger root in 150 minutes, while a multi-mode cavity-based HD took 90 minutes to yield 0.3 g of oil per 100 g ginger root. The findings from the single-mode cavity demonstrated that increasing the power and reducing the exposure time did not cause any degradation of the oil components. Mazidi et al. (2012) observed that, when compared to traditional HD, the MAHD technique enhances the extraction rate by rapidly elevating the temperature and hastening the rupture of the EO glands in black zira. In their investigation, Golmakani and Rezaei (2008) asserted that MAHD outperformed traditional HD in both energy conservation and extraction time, achieving extraction from *Thymus vulgaris* L. within 75 minutes as opposed to the 4 hours needed by HD. The MAHD method can be utilized both in a laboratory environment and at an industrial level (Kusuma and Mahfud, 2017; Drinić et al., 2021).

6.4.1.1 Vacuum MW Hydro-Distillation

The concept of vacuum MW hydro-distillation was developed to prevent the occurrence of hydrolysis (Li et al., 2014). This technique efficiently extracts compounds from plant material by employing selective MW heating, sequential vacuum application, and a suitable solvent (Chemat et al., 2013). The sample is positioned in an MW-compatible extraction vessel, where water is added to rehydrate the dry material. The choice of the solvent is determined by the properties of the compounds being extracted. MW irradiation generates rapid and uniform heating, facilitating the liberation of the target substances from the plant material into the solvent. The vacuum system assists in reducing the pressure inside the vessel. The decreased pressure enhances the solvent's boiling point, facilitating its vaporization at lower temperatures. The vaporized solvent, along with the extracted compounds, is collected in a separate container. Lowering the pressure to a range of 100 to 200 mbar facilitated the vaporization of the azeotropic mixture of water and volatile oil from the sample (Chemat et al., 2013; Movaliya, 2017). Moreover, conducting refluxing at low temperatures promotes adequate mixing and extraction of thermally sensitive compounds (Bagade and Patil, 2021). Yang et al. (2012) utilized an MW system equipped with a vacuum setup to enhance the yield of fresh garlic EO, employing a 95% wt. aqueous ethanol solution as the solvent for extraction. The study revealed that the yield from the MAE was 2.5 times more than what was obtained through the traditional technique. The utilization of both MW heating and decreased pressure allows for quicker extraction times in contrast to conventional extraction methods.

6.4.2 Ultrasonic MW-Assisted Extraction

This extraction method combines ultrasonic waves with MAE. The extraction configuration is a modified version of the coaxial MW hydro-distillation (MWHD) apparatus, utilizing a 1-L three-necked flask that accommodates a coaxial MW antenna, an ultrasonic metallic horn, and a Clevenger condenser, as shown in Figure 6.2. Both the ultrasonic horn and the MW antenna are submerged in the liquid phase and positioned at right angles to one another (González-Rivera et al.,

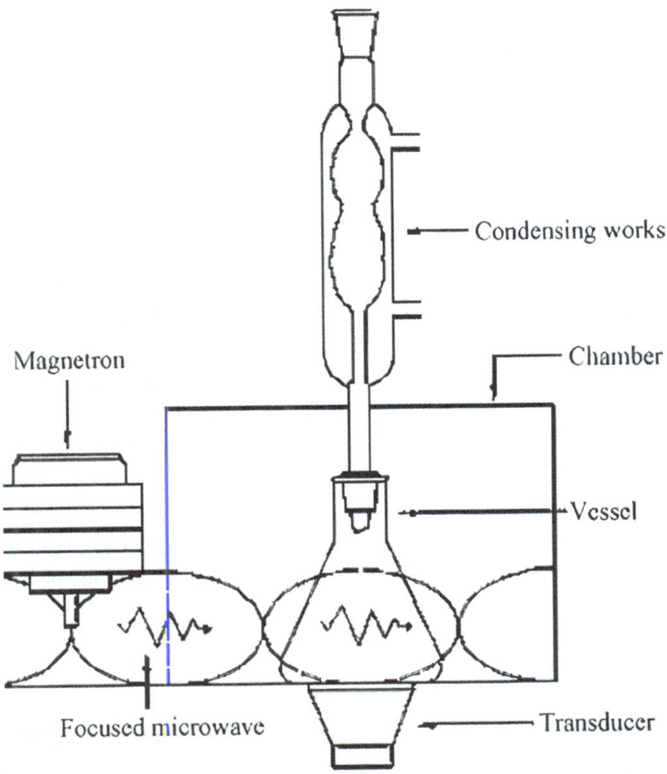

FIGURE 6.2 Schematic diagram of UMAE apparatus.

Source: Copyright obtained from Elsevier (Chen et al., 2010).

2016). This ensures optimal performance, maintains safe operational conditions, and avoids any interference when employing the two energy sources simultaneously. The setup generates substantial momentum and energy, enabling the solvent to permeate the cells of the sample by inducing their rupture. Consequently, the bioactive components are released more efficiently and rapidly from the plant tissue during this extraction process, resulting in lower solvent usage and accelerated extraction (González-Rivera et al., 2016; Bagade and Patil, 2021). González-Rivera et al. (2016) introduced an innovative approach to extracting EO from citrus peels by combining MAHD with US. They compared their findings with those from coaxial solvent-less MW-assisted extraction (SMWAE), coaxial MWHD, and conventional HD (CH). The achieved maximum yields were 1.16%, 1.53%, 1.57%, and 1.55% (w/w) within 5, 60, 76, and 155 minutes for SMWAE, US-MWHD, MWHD, and CH, respectively. The energy expenditure for each gram of EO extracted through HD was 0.83 kWh/g. By utilizing MWHD, this value decreases to 0.31 kWh/g, and with the application of the US-MWHD technique, it further decreases to 0.28 kWh/g. This signifies energy conservation of over 60% when compared to HD. Lianfu and Zelong (2008) utilized a combination of an open MW (800 W, 2450 MHz) and an ultrasonic transducer (50 W, 40 KHz) for lycopene extraction from tomato paste, and they compared this method with UAE. Response surface methodology (RSM) was employed to identify the ideal parameters for ultrasonic MW-assisted extraction (UMAE), yielding the following optimized settings: an MW power of 98 W, extraction duration of 367 seconds, and solvent-to-tomato paste ratio of 10.6:1 with a corresponding yield of 97.4% lycopene. In contrast, UAE under optimal conditions (temperature of 86.4°C, extraction period of 29.1 minutes, and solvent to tomato paste ratio of 8.0:1) yielded 89.4% lycopene. The assessment of these approaches revealed that UMAE outperformed UAE, indicating its potential as a more promising extraction method for the future. Wang et al. (2018) conducted a study comparing the yields of EO and free radical scavenging activities of white and black peppers through UMAE, UAE, and MAE techniques. The research found that UMAE, leveraging acoustic cavitation along with rapid heating through ultrasonic and MW waves, notably enhanced EO extraction. This improvement was evidenced by shorter extraction times, increased EO yields, and higher quality attributes when compared to both MAE and UAE methods. In their study, Chen et al. (2010) validated the significant potential and effectiveness of UMAE in comparison to the conventional method of hot water extraction for obtaining polysaccharides from *Inonotus obliquus*.

6.4.3 MW Steam Distillation

The technique is built upon the conventional SD principle, where the extraction reactor exclusively employs MW radiation. The cooling system and the component responsible for collecting essences are positioned outside the MW oven (Sahraoui et al., 2008; Chemat et al., 2013). The arrangement of this setup involves placing the plant material above the water within the MW oven as shown

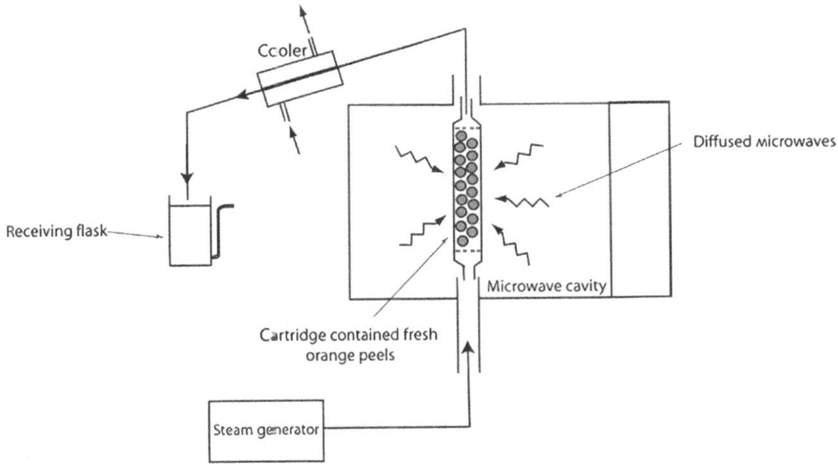

FIGURE 6.3 MW steam distillation.

Source: Copyright obtained from Elsevier (Sahraoui et al., 2011).

in Figure 6.3, with a Teflon grid providing separation (Bagade and Patil, 2021). The steam serves as the sole carrier gas that interacts with the sample, leading to its evaporation and guiding the EO through the condenser into a collection vessel. Sahraoui et al. (2008) utilized this method to extract EO from lavender flowers (*Lavandula angustifolia* Mill.). With the MSD technique, lavender flowers can achieve comparable yields to those obtained through the SD method, but in just 6 minutes instead of the 30 minutes required by SD. An essential characteristic of the MSD reactor is its ability to apply MW radiation to lavender flowers selectively, generating localized "hot spots" within the process. The effectiveness of this method far surpasses that of the traditional SD process, considering its quick distillation times, cost-effectiveness, energy efficiency, and the cleanliness it offers during the procedure.

6.4.4 MW Hydrodiffusion and Gravity

This innovative "upside down" MW alembic represents an environmentally friendly extraction approach that integrates MW heating with the force of gravity at standard atmospheric pressure as shown in Figure 6.4 (Vian et al., 2008; Bousbia et al., 2009a; Chemat et al., 2013; Calinescu et al., 2017). In this approach, fresh plant sample is directly introduced into an MW reactor, eliminating the need for additional solvents or water (Movaliya, 2017). MWs generate heat within the matrix which causes the plant cells to expand, resulting in the breaking of glands and oil-containing structures. The heat generated by MWs enables the liberation of EO and water from the plant matrix through the hydro-diffusion phenomenon. The resulting extract, comprising both EO and water, flows downward

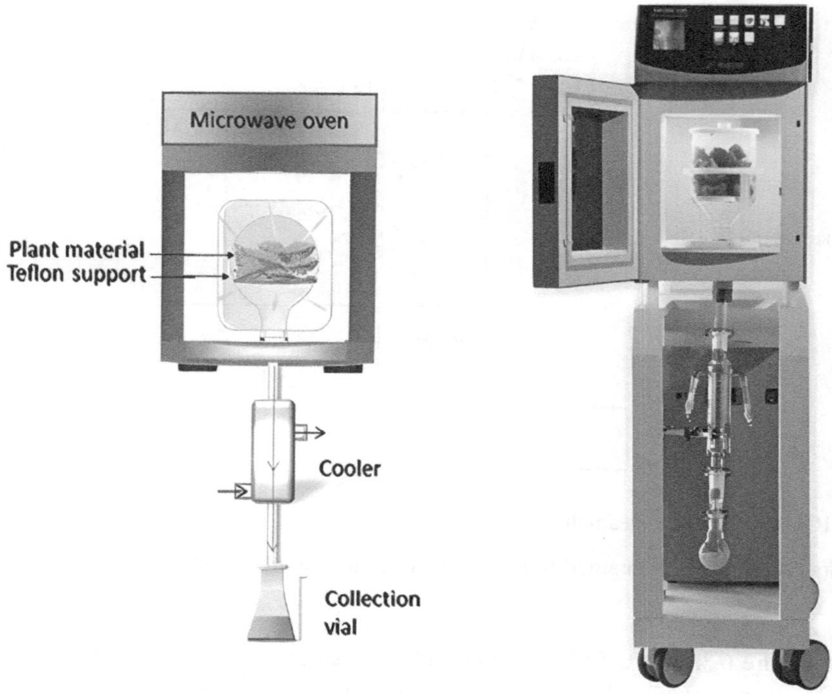

FIGURE 6.4 MW hydro-diffusion and gravity.

Source: Copyright obtained from Elsevier (Li et al., 2013).

under the influence of gravity via the perforated Pyrex disk in the MW reactor (Calinescu et al., 2017). It then moves through a spiral condenser situated outside the MW cavity, where EOs are recovered via condensation of the distillate (Ciriminna et al., 2016). MHG ensure uniform heat distribution within the irradiated medium. The MHG system mainly consists of four components: a reactor for the processed sample, an MW oven, a cooling system, and an essence vessel used for retrieving EO (Chemat et al., 2013). It is noteworthy that this method allows for EO extraction without requiring energy-intensive processes such as distillation and evaporation, which are among the most demanding operations in terms of energy consumption (Vian et al., 2008; Bousbia et al., 2009a). This method has been applied for extracting EOs from spearmint (*Mentha spicata* L.) and pennyroyal (*Mentha pulegium* L.). Using MHG, extraction duration of 20 minutes produces outcomes comparable to those obtained from 90 minutes of HD (Vian et al., 2008). Through their research, Bousbia et al. (2009a) showcased that by utilizing MHG, EOs from citrus fruits could be obtained in just 15 minutes, achieving results comparable to those achieved through a 180-minute HD process. The MHG technique offers the capacity to closely replicate the natural aroma present in citrus fruit EOs, similar to CP, and outperforms hydro-distilled EOs. Bousbia et al. (2009b) achieved a 0.33% EO yield from fresh *Rosmarinus officinalis* leaves using an optimal MW power of 500 W and an isolation time of 15 minutes with MHG, which was a good alternative to the 0.35% yield obtained after 180 minutes using HD. This green extraction technique presents benefits in terms of cost reduction and substantial time and energy savings (Vian et al., 2008; Bousbia et al., 2009a, 2009b; Chemat et al., 2013).

6.4.5 Solvent-Free MW Extraction

SFME is an innovative approach that combines MW irradiation and dry distillation as depicted in Figure 6.5. It is conducted under normal atmospheric pressure and does not require the use of solvents or water (Lucchesi et al., 2004; Uysal et al., 2010; Filly et al., 2014). It was created and officially patented in 2004 by Chemat et al. (2004). The device consists of an MW oven featuring a chamber specifically designed to accommodate biological material. Furthermore, it includes a condensation chamber located above and external to the MW chamber, connected to it through a connecting channel. The SFME method distinguishes itself from modified MAE that employs organic solvents and modified HD, which necessitates a

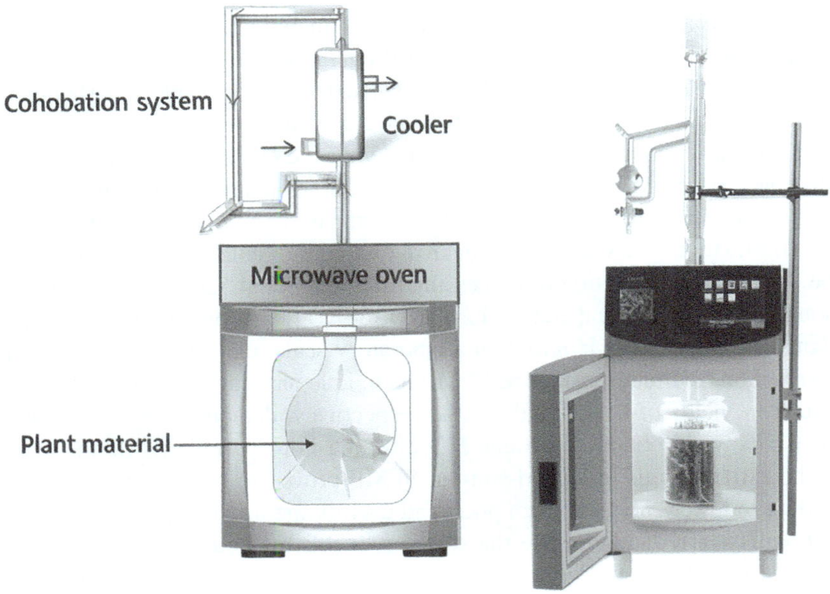

FIGURE 6.5 Solvent-free MW extraction.

Source: Copyright obtained from Elsevier (Li et al., 2013).

significant volume of water. The process involves placing the biological material into a solvent-free MW chamber, subjecting it to MW radiation until the natural substance is liberated, transferring the extract from the MW chamber to a condensation chamber through convection, cooling the substance until condensation occurs, and ultimately transferring the condensed natural substance from the condensation chamber (Chemat et al., 2004). González-Rivera et al. (2016) developed a coaxial solvent-less MW-assisted extraction approach (SMWAE) for extracting aromatic compounds from citrus peels and conducted a comparative analysis with results obtained from simultaneous US coaxial MW-assisted HD (US-MWHD), coaxial MWHD, and CH methods. The highest yields achieved were 1.16%, 1.53%, 1.57%, and 1.55% (w/w) within 5, 60, 76, and 155 minutes for SMWAE, US-MWHD, MWHD, and CH, respectively. Comparing the energy consumption throughout various stages of the orange peel bioconversion process for each unit of product mass, the SMWAE method (0.03 kWh/g) emerged as the most favorable option, being approximately 27 times more efficient than the CH method. This was made possible by the coaxial antenna's *in-situ* MW irradiation, resulting in a combination of selective and uniform rapid heating from the inner to the outer layers. Achieving higher efficiency in SFME is possible with samples that exhibit a higher dielectric loss (indicating high water content). This enhanced efficiency results from the strong interaction between MWs and the physiological water, which contains salts and nutrients within the samples (Filly et al., 2014). In their research, Lucchesi et al. (2004) explored the effectiveness of the SFME approach, contrasting it with HD, in extracting EOs from the aerial components of three aromatic herbs: basil (*Ocimum basilicum* L.), garden mint (*Mentha crispa* L.), and thyme (*Thymus vulgaris* L.). The results clearly demonstrate that utilizing the SFME technique for a 30-minute extraction period produces results comparable to those attained with the HD method after 4.5 hours. Moreover, the SFME method presents a substantial advantage in terms of energy cost reduction, requiring only 0.25 kWh compared to the 4.5 kWh needed for the HD method. This significant decrease in energy consumption is a result of the considerable reduction in processing time. Filly et al. (2014) made a comparison between SFME and conventional HD for EO recovery from fresh rosemary plants (*Rosmarinus officinalis*) on a laboratory scale and pilot scale. Extraction duration of 30 minutes using the SFME technique on both bench and pilot scales results in yields of 0.54% and 0.50%, respectively, demonstrating similarity to the 0.57% yield obtained through HD after a 2-hour period. While the quality of the extracted oils remained similar, SFME proved to be more time-efficient, leading to lower energy consumption and overall operational costs. The cost and CO_2 emission reduction make the scale-up of this method highly attractive. As a result, the SFME approach yields valuable EOs, shorter extraction times, and significant energy efficiency. Research findings also indicate that EOs obtained through SFME show elevated quantities of oxidized substances and reduced levels of monoterpene hydrocarbons in contrast to those obtained through HD and UAE techniques (Lucchesi et al., 2004; Uysal et al., 2010; Filly et al., 2014).

6.4.6 Compressed Air MW Distillation

This technique was introduced by Craveiro et al. (1989) for extracting EO from *Lippia sidoides* Cham leaves. This process entails the continuous injection of compressed air into the extractor containing the plant material submerged in water, concurrently applying MW heating. The condensation and separation of water and EOs occur outside the MW reactor. The outcomes demonstrated that compressed air MW distillation (CAMD) achieved completion within 5 minutes, in contrast to the 60–90 minutes required for traditional SD. Notably, there was no variance in both the quantitative and qualitative aspects of the extracted substances. The synergistic effect of MW heating along with the forced convection of compressed air leads to rapid and effective extraction of the desired elements from the plant material. As a result, CAMD has the potential to decrease energy consumption.

6.4.7 Pressurized SFME

Pressurized SFME (PSFME) is achieved by subjecting the sample to controlled pressure while applying MW irradiation (Michel et al., 2011). It relies on the absorption of MW energy by internal water within the plant material, which subsequently acts as the heat source for the solvent–matrix solution. This interaction induces internal heating within the biomaterial, resulting in heightened pressure within plant cells. Consequently, this elevated pressure results in the breaking of cell walls, facilitating the liberation of analytes. For the PSFME process, as set up in Figure 6.6, it is necessary to use fresh or frozen plant material that possesses adequate moisture content. In a study by Michel et al. (2011), they examined the efficacy of PSFME in extracting antioxidants from *Hippophaë rhamnoides* L. berries, comparing it with conventional extraction techniques. The results revealed that PSFME enhanced the movement of molecules from the matrix into the solvent, leading to increased release and solubility of compounds. Notably, PSFME generated a potent and abundant extract in terms of phenolic content, encompassing compounds like quercetin and isorhamnetin that remained unextracted through other methods.

6.5 VARIABLES AFFECTING MAE EFFICIENCY

Several variables, including the solvent type and composition, extraction duration, sample characteristics, and MW power, influence the process of MAE. The choice of these parameters will be determined by the solubility, volatility, and stability of the desired compounds, as well as their interaction with other coexisting substances (Desai et al., 2010). Moreover, the bioactive properties of EOs are derived from their chemical makeup, which is shaped by the distinct plant species, maturity, harvest time, postharvest handling prior to isolation, prevailing climatic conditions, and extraction techniques (Calinescu et al., 2017; Drinić et al., 2021).

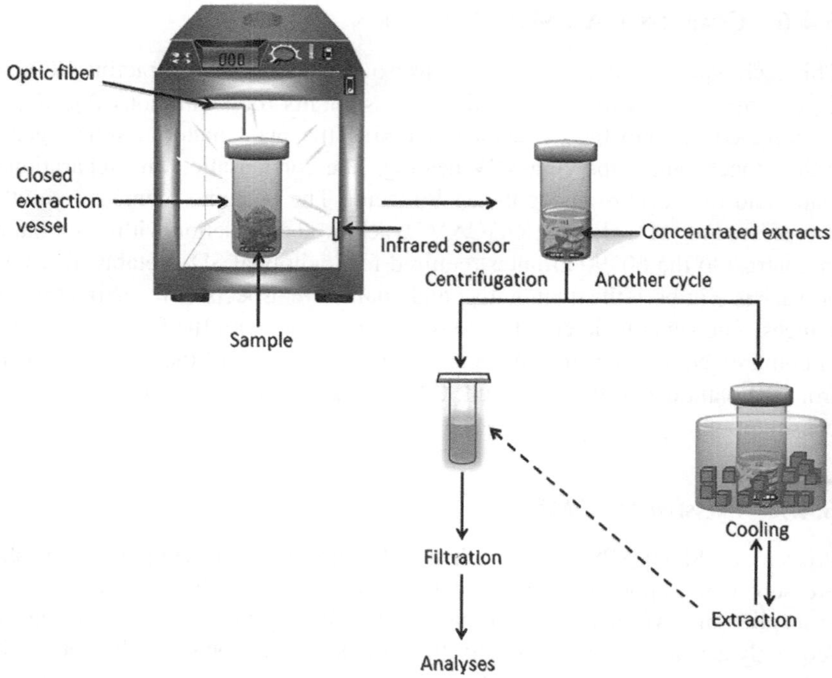

FIGURE 6.6 Pressurized solvent-free MW extraction.

Source: Copyright obtained from Elsevier (Li et al., 2013).

6.5.1 MW Power

The correlation between MW power and temperature is closely interrelated (Bagade and Patil, 2021). As the MW power is elevated to a certain threshold, more energy is delivered to the material, resulting in an increase in temperature that directly impacts the yields obtained during the extraction process (Routray and Orsat, 2012). Studies conducted by different researchers have affirmed that as power irradiation increases, the mass transfer ratio rises until it reaches a specific threshold, ultimately leading to a higher extraction yield (Abedi et al., 2017; Liu et al., 2018a). The increased EO yield at higher MW power has been attributed to a synergistic effect resulting from the temperature and mass gradient, both working in unison. However, when the MW power is decreased, the heating ability diminishes, leading to a lower extraction yield (Drinić et al., 2020). Liu et al. (2018a) observed that with an increase in MW power from 400 W, the mass transfer rate accelerates, resulting in a higher extraction yield of EOs obtained from lavender. Nevertheless, when the power exceeded 500 W, there was a decrease in the yield. High temperatures promote interactions between molecules in the solvent, resulting in higher molecular mobility and consequently boosting solubility

(Routray and Orsat, 2012). Increasing the temperature allows solvents to penetrate the cell wall more quickly, accelerates the liberation of the desired molecule from the plant material, and improves overall extraction efficiency (Hou et al., 2019). Moreover, high temperatures lead to a reduction in solvent viscosity, resulting in increased mobility and solubility. This, in turn, enhances the extraction efficiency.

The extraction efficiency improves as the temperature rises toward the optimal temperature, which is determined by the stability of the target compounds. Therefore, selecting the appropriate MW power is of utmost importance since excessively high power levels can cause the thermal decomposition of volatile substances (Akhbari et al., 2018; Liu et al., 2018a; Drinić et al., 2020). At excessively high temperatures, there can be a reduction in extraction yield due to the instability of the compounds (Kusuma and Mahfud, 2017). Using temperatures near the solvent's boiling point can accelerate the interaction between molecules and improve solute solubility in the solution, leading to quicker extraction times (Deepak et al., 2013; Martínez-Abad et al., 2020).

6.5.2 Sample Characteristics

The size of the particles employed has a notable impact on the EO recovery through the MAE technique. Reduced particle sizes lead to increased surface area relative to volume, enhancing synergy between the plant material and the solvent. This interaction improves extraction efficiency and yields. Moreover, smaller particles enable more efficient absorption of MW energy into the plant material, leading to enhanced heating and EO extraction (Delazar et al., 2012; Desai and Parikh, 2012). The presence of moisture within the matrix can significantly increase MW absorption and promote heating by influencing the solvent's polarity (Bagade and Patil, 2021). In some cases, a finer particle size of the matrix can create challenges when attempting to separate the matrix from the solvent following the extraction process. To address this concern, an extra step involving the application of centrifugation and filtration can be employed as a solution (Mandal et al., 2007).

6.5.3 Extraction Time

The extraction time significantly influences the efficiency, yield, sample integrity, selectivity, and energy consumption of the extraction process. Generally, longer extraction times result in enhanced extraction efficiencies. By increasing the extraction time, the solvent can better penetrate and diffuse deeper into the plant matrix, enabling the liberation of EO constituents from the sample. Various studies on MAE have consistently observed two distinct extraction phases, irrespective of the plant material or solvent polarity. The first phase, known as the washing stage, involves a swift rise in the level of extractable compounds in the solvent. The second phase, referred to as the diffusion step, is characterized by a gradual increment in the level of extractable compounds within the solvent, which intensifies as the extraction procedure advances (Sahraoui et al., 2008; Filly et al., 2014; Lee et al., 2016; Kusuma and Mahfud, 2017; Franco-Vega et al., 2019). In the research conducted by Liu et al. (2018a), it was noted that extending the time for extraction while keeping the power level constant resulted in a rise in the lavender EO production, though this increase was not observed until the extraction time surpassed 38 minutes. Lee et al. (2016) observed that conventional hydrothermal extraction resulted in extraction peaks of 18% after 120 minutes, whereas MAE achieved significantly higher extraction peaks of 48% within a shorter time frame of just 10 minutes. In the study conducted by Kusuma and Mahfud (2017) regarding EO extraction from Patchouli (*Pogostemon cablin*) leaves, it was observed that the EO droplets appeared after 57 minutes in the HD method and only 6 minutes in the MAHD method. This notable difference was attributed to the application of heat transfer mechanisms in the samples through irradiation, conduction, and convection in the MAHD process. The MW's ability to facilitate more efficient heat transfer is primarily responsible for its superior extraction rates. This efficiency leads to shorter extraction times. Nevertheless, there exists an optimal duration for extraction beyond which additional processing might not substantially enhance the yield. If the extraction time is excessively prolonged, degradation of the target compounds may occur due to the increased exposure to MW energy (Hou et al., 2019). This could result in alterations to the chemical composition of the extracted compounds, affecting their quality and causing a subsequent decrease in yield. In addition, the time of exposure is influenced by the dielectric characteristics of the solvent (Kusuma and Mahfud, 2017; Bagade and Patil, 2021). Due to its high dielectric constant, water prompts immediate heating in the course of the extraction procedure (Drinić et al., 2020). Utilizing an MW-absorbing solvent such as water carries the potential risk of thermolabile components in the plant matrix undergoing degradation. By reducing the extraction time, it becomes possible to preserve a greater quantity of thermolabile components in the end product (Calinescu et al., 2014). Therefore, it is vital to establish a harmonious equilibrium in the extraction yield and the stability of the constituents to ensure a meaningful MAE procedure. Also, different components in the plant material have varying susceptibilities to MW energy and solvent interactions. Prolonged extraction times may extract not only the desired EO compounds but also undesirable components, affecting the overall process selectivity. The duration of extraction also influences consumption of energy and CO_2 emissions, as longer extraction times necessitate more energy input, which can have implications for the overall cost and sustainability of the extraction method (Drinić et al., 2021). Therefore, it is essential to find the ideal extraction duration for optimal output of the desired compounds, which will help minimize energy consumption while maintaining satisfactory efficiency of the extraction procedure. Determining the appropriate extraction time requires considering the nature of the sample, the target compounds, and their stability under MW conditions.

6.5.4 Solvent Selection

In any MAE process, the choice of a suitable solvent and the establishment of an optimal solvent-to-solid ratio are of utmost importance. A wide range of solvents with different polarities, typically used in traditional extraction techniques (such as diethyl ether, ethyl acetate, acetone, *n*-hexane, and ethanol), can be effectively employed in MAE. Selecting an appropriate solvent for MAE relies on factors such as the solubility of the desired compound, the MW-absorbing characteristics of the solvent, and the interplay between the solvent and the matrix (Delazar et al., 2012; Bagade and Patil, 2021). The solvent typically interacts with the cell wall of the biological sample, permeating it in order to access the constituents. The ideal solvent should demonstrate a higher degree of selectivity towards the target analyte when compared to the other constituents in the matrix (Routray and Orsat, 2012). Additionally, it should possess a strong ability to absorb MW energy, facilitating rapid heating and enhancing the extraction process. Moreover, the solvent's dielectric susceptibility and the kinetics of mass transfer are significant considerations that influence the selection of solvent during the MAE process (Lee et al., 2016; Martínez-Abad et al., 2020). Usually, in cases where there is a high dielectric constant and dielectric loss, the solvent has an increased capacity for absorbing MW energy (Routray and Orsat, 2012). This, in turn, can accelerate the rate at which the solvent heats up with respect to the plant material. Solvents such as water possess excellent MW energy absorption capabilities, promoting the enhanced liberation of EOs (Liu et al., 2018a). Polar solvents like ethanol and methanol exhibit lower dielectric constant values, resulting in less MW absorption compared to water. However, due to their increased dissipation factor values, the overall efficiency of both solvents remains higher. In contrast, nonpolar solvents like hexane and chloroform, which have lower polarities, do not generate heat as they are transparent to MWs (Tatke and Jaiswal, 2011). MW extraction can be conducted using both polar and nonpolar solvents (Routray and Orsat, 2012). Extraction yields can be increased by immersing the plant matrix in solvents that have high dissipation factor values. The efficiency of the solvent during the entire process is influenced by the amount of plant matrix present (Bagade and Patil, 2021). Adequate solvent volume is essential to fully submerge the plant matrix throughout the entire irradiation process, particularly if the matrix is prone to swelling during the extraction procedure. By simultaneously minimizing the use of excessive solvent, one can effectively reduce process delays during extraction (Franco-Vega et al., 2019).

In conventional extraction methods, it is commonly observed that higher recovery rates are achieved by increasing the solvent volume (Liu et al., 2018a). Nevertheless, when employing MAE, an increased solvent volume could lead to reduced recovery rates and wastage of solvent (Delazar et al., 2012; Hou et al., 2019). Typically, the quantity of solvent required for a single sample falls between 10 and 30 ml (Delazar et al., 2012; Chutia and Mahanta, 2021). Some researchers have suggested that achieving optimal extraction yields and isolating specific active compounds from a plant sample can be accomplished by using blends of solvents with varying MW absorption capabilities (Desai et al., 2010; Delazar et al., 2012; Routray and Orsat, 2012; Bagade and Patil, 2021).

6.6 ECONOMY, ENERGY, AND ENVIRONMENTAL IMPLICATIONS OF MAE

In the face of rising energy costs and as the drive for CO_2 emission reduction intensifies, industries face the task of seeking innovative technologies that can lower energy consumption and comply with emission regulations while also achieving cost efficiencies and enhanced quality standards. In today's industrial production context, such key terms have gained paramount importance. A key benefit of MAE over conventional methods is the cost reduction, which is directly linked to the decreased energy requirements for achieving the desired analyte. This cost-effectiveness is clearly attributed to the reduced extraction time (Vian et al., 2008; Kusuma and Mahfud, 2017; Elyemni et al., 2019; Franco-Vega et al., 2019). The energy requirement for carrying out the extraction methods, determined using the maximum power consumption, is evaluated using a wattmeter at both the MW power source and steam generation sources. In terms of environmental influence, the estimated amount of CO_2 released into the atmosphere is determined using the following assumptions: for every 1 kWh obtained from fuel, approximately 800 g of CO_2 is emitted into the air through the burning of fossil fuels (Ferhat et al., 2007; Sahraoui et al., 2008; Vian et al., 2008).

According to Golmakani and Moayyedi (2015), the EO extraction of Citrus limon peel required 0.67 kWh of energy using HD, whereas both MAHD and SFME consumed only 0.30 kWh. In terms of environmental pollution, the measured CO_2 emissions were more pronounced with HD (0.533 kg CO_2), compared to the levels associated with MAHD and SFME (0.240 kg CO_2). In their study, Jeyaratnam et al. (2016) noted that the energy needed for extracting EOs from *Cinnamomum cassia* bark was 1.0 kWh using the HD method, while the MAHD method required 0.4 kWh. When considering the environmental implications, a larger quantity of CO_2 is emitted into the environment with HD (779 g CO_2/g oil) compared to MAHD (330 g CO_2/g oil). In the study conducted by Kusuma and Mahfud (2017), the relative electric consumption for producing 1 ml of EO from patchouli (*Pogostemon cablin*) leaves was 13.8670 kWh/ml using conventional HD and 2.3134 kWh/ml using MAHD. In terms of environmental consequences, the research highlighted a substantial difference in the relative quantity of CO_2 discharged into the air for each 1 ml of EO. Specifically, the emissions were notably higher for conventional HD (11.0936 kg CO_2/g EO) in comparison to MAHD (1.8507 kg CO_2/g EO). In their study, Vian et al. (2008) found that the energy demands for carrying out the two extraction processes on spearmint and pennyroyal EOs were 3 kWh (90 min) for HD and 0.2 kWh (15 min) for MHG, respectively. The computed emission of CO_2 into the atmosphere was notably higher for HD (2400 g CO_2) compared to MHG (160 g CO_2). Moreover, MHG plays a role in pollution prevention by having the potential to save up to 90% of energy, thus contributing to the reduction of greenhouse gas emissions. Sahraoui et al. (2008) observed that traditional SD required an extraction duration of 30 minutes. However, the MSD method achieved the same EO recovery in just 6 minutes. The energy demands for carrying out the distillation processes are 0.3 kWh/g of EO for SD and 0.05 kWh/g of EO for MSD. In terms of environmental repercussions, the calculated quantity of CO_2 emitted into the atmosphere is greater for SD (240 g CO_2/g of EO), compared to MSD (40 g CO_2/g of EO). For EO extraction from *Cinnamomum camphora*, SFME outperformed conventional HD across various parameters. Specifically, SFME exhibited higher extraction effectiveness (3.51 ± 0.12% within 23 minutes in contrast to 3.35 ± 0.10% within 240 minutes), lower energy usage (0.22 kW h as opposed to 4 kW h), and reduced carbon footprint (177.87 g CO_2 instead of 3200 g CO_2), as evidenced in the research conducted by Liu et al. (2018b). These findings clearly demonstrate a significant reduction in energy consumption, extraction costs, and carbon emissions, thus affirming MAE as an economically viable technique for implementation on an industrial scale (Ciriminna et al., 2016).

6.7 FUTURE PROSPECTS

As researchers continue to explore potential avenues for further advancements in MAE methodologies, several key factors come to light. Optimizing key parameters has the potential to improve the effectiveness of MAE, leading to higher yields of EO while concurrently reducing extraction times, energy consumption, and overall production costs. By adjusting the MW power, extraction time, solvent choice, and sample size, one can attain an optimal balance of efficiency and effectiveness, ensuring the best possible results in the extraction process. The optimization of process parameters, including MW power, liquid-to-solid ratio, and MW time for the maximum yield of lavender EO through the MAHD method, was conducted using RSM with a Box–Behnken design (BBD), as outlined by (Liu et al., 2018a). Hou et al. (2019) employed the BBD to optimize the extraction of EO from *Pinus koraiensis*, achieving a yield of 7.76 mg/g.

Although MAE has demonstrated remarkable potential in laboratory environments, an increasing interest exists in adapting this technology for industrial applications. The feasibility and economic significance of upscaling the MW-based extraction technique to an industrial level are undoubtedly substantial. According to Ciriminna et al. (2016), the key consideration to take into account when transitioning to larger scales is the thickness of the sample and the frequency of electromagnetic radiation. Achieving deeper penetration of electromagnetic waves in large-scale extraction may require a lower frequency (0.915 GHz); however, this could lead to slower and uneven sample heating. On the other hand, opting for a higher frequency (2.45 GHz) is preferable, but it comes with the limitation of a maximum sample thickness of 2 cm due to the shorter wavelength's penetration capabilities (Ciriminna et al., 2016). Bradshaw et al. (1998) indicated that hybrid heating systems prove highly beneficial in addressing the issue of temperature uniformity.

The exploration of innovative extraction configurations, integrating MAE with other methods like UAE or supercritical fluid extraction, is also garnering attention. These synergistic techniques seek to leverage the benefits of multiple methods, resulting in enhanced efficiency and selectivity in the extraction process. González-Rivera et al. (2016) developed an integrated extraction system, allowing simultaneous irradiation of MWs and US for the HD of EO from organic waste orange peels. This technology showcased remarkable results boasting over 60% energy savings compared to traditional HD. Beyond its efficiency, the system demonstrated substantial improvements in yield, product quality, economic viability, and environmental sustainability. Cravotto et al. (2008) achieved higher extraction efficiency at moderate cost with simultaneous US/MW irradiation, surpassing the performance of conventional methods for extracting soybean-germ oil.

6.8 CONCLUSION

MAE is significant as an innovative substitute for traditional extraction techniques in obtaining EOs. The remarkable attributes of this environmentally friendly approach, including enhanced extraction efficiency, reduced extraction time, minimized energy consumption, expedited energy transfer, reduced solvent usage, and simplified handling while maintaining a high-quality extract, have attracted considerable interest from researchers, scientists, and industries. Its versatility in accommodating varying sample sizes, temperatures, pressures, and solvent volumes, coupled with its ability to rapidly deactivate the heat source, enhances its applicability to diverse botanical materials. While recognizing the advantages of MAE, it is important to take into account the associated limitations. These include the possibility of heat-sensitive bioactive components undergoing degradation and the requirement for supplementary clean-up procedures. Addressing these challenges will require further research aimed at optimizing the MAE process, particularly in terms of parameter control and regulation for various plant matrices and target compounds. Future advancements could include the integration of MAE with other innovative techniques, like UAE or supercritical fluid extraction, and expanding its applicability to industrial scales. Such developments could open up new pathways for efficient and sustainable EO extraction.

REFERENCES

Abedi, A.-S., Rismanchi, M., Shahdoostkhany, M., Mohammadi, A. and Mortazavian, A. M. (2017). Microwave-Assisted Extraction of Nigella Sativa L. Essential Oil and Evaluation of Its Antioxidant Activity. *Journal of Food Science and Technology,* 54: 3779–3790.

Akhbari, M., Masoum, S., Aghababaei, F. and Hamedi, S. (2018). Optimization of Microwave Assisted Extraction of Essential Oils from Iranian Rosmarinus Officinalis L. Using Rsm. *Journal of Food Science and Technology,* 55: 2197–2207.

Bagade, S. B. and Patil, M. (2021). Recent Advances in Microwave Assisted Extraction of Bioactive Compounds from Complex Herbal Samples: A Review. *Critical Reviews in Analytical Chemistry,* 51: 138–149.

Bale, A. and Shinde, N. (2013). Research Article Microwave Assisted Extraction of Essential Oil from Lemon Leaves. *International Journal of Recent Scientific Research,* 4: 1414–1417.

Bousbia, N., Vian, M. A., Ferhat, M. A., Meklati, B. Y. and Chemat, F. (2009a). A New Process for Extraction of Essential Oil from Citrus Peels: Microwave Hydrodiffusion and Gravity. *Journal of Food Engineering,* 90: 409–413.

Bousbia, N., Vian, M. A., Ferhat, M. A., Petitcolas, E., Meklati, B. Y. and Chemat, F. (2009b). Comparison of Two Isolation Methods for Essential Oil from Rosemary Leaves: Hydrodistillation and Microwave Hydrodiffusion and Gravity. *Food Chemistry,* 114: 355–362.

Bustamante, J., van Stempvoort, S., García-Gallarreta, M., Houghton, J. A., Briers, H. K., Budarin, V. L., . . . & Clark, J. H. (2016). Microwave assisted hydro-distillation of essential oils from wet citrus peel waste. *Journal of Cleaner Production*, 137, 598–605.

Bradshaw, S., Van Wyk, E.J. and De Swardt, J. (1998). Microwave Heating Principles and the Application to the Regeneration of Granular Activated Carbon. *Journal of the Southern African Institute of Mining and Metallurgy,* 98: 201–210.

Calinescu, I., Asofiei, I., Gavrila, A. I., Trifan, A., Ighigeanu, D., Martin, D., Matei, C. and Buleandra, M. (2017). Integrating Microwave-Assisted Extraction of Essential Oils and Polyphenols from Rosemary and Thyme Leaves. *Chemical Engineering Communications,* 204: 965–973.

Calinescu, I., Gavrila, A., Ivopol, M., Ivopol, G., Popescu, M. and Mircioaga, N. (2014). Microwave Assisted Extraction of Essential Oils from Enzymatically Pretreated Lavender (Lavandula Angustifolia Miller). *Open Chemistry,* 12: 829–836.

Cardoso-Ugarte, G. A., Juárez-Becerra, G. P., SosaMorales, M. E. and López-Malo, A. (2013). Microwave-Assisted Extraction of Essential Oils from Herbs. *Journal of Microwave Power and Electromagnetic Energy,* 47: 63–72.

Chaturvedi, A. K. (2018). Extraction of Neutraceuticals from Plants by Microwave Assisted Extraction. *Systematic Reviews in Pharmacy,* 9: 31–35.

Chemat, F., Abert-Vian, M. and Fernandez, X. (2013). Microwave-Assisted Extraction of Essential Oils and Aromas. *Microwave-Assisted Extraction for Bioactive Compounds: Theory and Practice,* 53–68.

Chemat, F., Lucchesi, M. E. and Smadia, J. (2004). *Solvent-Free Microwave Extraction of Volatile Natural Substances*. US Patent App. 10/751,988.

Chen, Y., Gu, X., Huang, S.-q., Li, J., Wang, X. and Tang, J. (2010). Optimization of Ultrasonic/Microwave Assisted Extraction (Umae) of Polysaccharides from Inonotus Obliquus and Evaluation of Its Anti-Tumor Activities. *International Journal of Biological Macromolecules,* 46: 429–435.

Chutia, H. and Mahanta, C. L. (2021). Green Ultrasound and Microwave Extraction of Carotenoids from Passion Fruit Peel Using Vegetable Oils as a Solvent: Optimization, Comparison, Kinetics, and Thermodynamic Studies. *Innovative Food Science & Emerging Technologies,* 67: 102547.

Ciriminna, R., Carnaroglio, D., Delisi, R., Arvati, S., Tamburino, A. and Pagliaro, M. (2016). Industrial Feasibility of Natural Products Extraction with Microwave Technology. *ChemistrySelect,* 1: 549–555.

Craveiro, A., Matos, F., Alencar, J. and Plumel, M. (1989). Microwave Oven Extraction of an Essential Oil. *Flavour and Fragrance Journal,* 4: 43–44.

Cravotto, G., Boffa, L., Mantegna, S., Perego, P., Avogadro, M. and Cintas, P. (2008). Improved Extraction of Vegetable Oils under High-Intensity Ultrasound and/or Microwaves. *Ultrasonics Sonochemistry,* 15: 898–902.

Datta, A. and Rakesh, V. (2013). Principles of Microwave Combination Heating. *Comprehensive Reviews in Food Science and Food Safety,* 12: 24–39.

Deepak, G., Mumtaj, S. and Prashant, S. (2013). Microwave-Assisted Extraction of Eucalyptus Citriodora Oil and Comparison with Conventional Hydro Distillation. *Middle East Journal of Scientific Research,* 16: 702–705.

Delazar, A., Nahar, L., Hamedeyazdan, S. and Sarker, S. D. (2012). Microwave-Assisted Extraction in Natural Products Isolation. *Natural Products Isolation,* 89–115. Humana Press.

Desai, M., Parikh, J. and Parikh, P. (2010). Extraction of Natural Products Using Microwaves as a Heat Source. *Separation & Purification Reviews,* 39: 1–32.

Desai, M. A. and Parikh, J. (2012). Microwave Assisted Extraction of Essential Oil from Cymbopogon Flexuosus (Steud.) Wats.: A Parametric and Comparative Study. *Separation Science and Technology,* 47: 1963–1970.

Drinić, Z., Jovanović, M., Pljevljakušic, D., Ćujić-Nikolić, N., Bigović, D. and Šavikin, K. (2021). Microwave-Assisted Extraction of Essential Oil from Ginger (*Zingiber Officinale* Rosc.). *Lekovite sirovine,* 41(1): 22–27.

Drinić, Z., Pljevljakušić, D., Živković, J., Bigović, D. and Šavikin, K. (2020). Microwave-Assisted Extraction of O. Vulgare L. Spp. Hirtum Essential Oil: Comparison with Conventional Hydro-Distillation. *Food and Bioproducts Processing,* 120: 158–165.

Elyemni, M., Louaste, B., Nechad, I., Elkamli, T., Bouia, A., Taleb, M., Chaouch, M. and Eloutassi, N. (2019). Extraction of Essential Oils of Rosmarinus Officinalis L. By Two Different Methods: Hydrodistillation and Microwave Assisted Hydrodistillation. *The Scientific World Journal,* 2019.

Ferhat, M. A., Meklati, B. Y. and Chemat, F. (2007). Comparison of Different Isolation Methods of Essential Oil from Citrus Fruits: Cold Pressing, Hydrodistillation and Microwave 'Dry'distillation. *Flavour and Fragrance Journal,* 22: 494–504.

Filly, A., Fernandez, X., Minuti, M., Visinoni, F., Cravotto, G. and Chemat, F. (2014). Solvent-Free Microwave Extraction of Essential Oil from Aromatic Herbs: From Laboratory to Pilot and Industrial Scale. *Food Chemistry,* 150: 193–198.

Fiorini, D., Scortichini, S., Bonacucina, G., Greco, N. G., Mazzara, E., Petrelli, R., Torresi, J., Maggi, F. and Cespi, M. (2020). Cannabidiol-Enriched Hemp Essential Oil Obtained by an Optimized Microwave-Assisted Extraction Using a Central Composite Design. *Industrial Crops and Products,* 154: 112688.

Franco-Vega, A., Ramírez-Corona, N., López-Malo, A. and Palou, E. (2019). Studying Microwave Assisted Extraction of Laurus Nobilis Essential Oil: Static and Dynamic Modeling. *Journal of Food Engineering,* 247: 1–8.

Golmakani, M.-T. and Rezaei, K. (2008). Comparison of Microwave-Assisted Hydrodistillation Withthe Traditional Hydrodistillation Method in the Extractionof Essential Oils from Thymus Vulgaris L. *Food Chemistry,* 109: 925–930.

Golmakani, M. T. and Moayyedi, M. (2015). Comparison of Heat and Mass Transfer of Different Microwave-Assisted Extraction Methods of Essential Oil from Citrus Limon (Lisbon Variety) Peel. *Food Science & Nutrition,* 3: 506–518.

González-Rivera, J., Spepi, A., Ferrari, C., Duce, C., Longo, I., Falconieri, D., Piras, A. and Tine, M. R. (2016). Novel Configurations for a Citrus Waste Based Biorefinery: From Solventless to Simultaneous Ultrasound and Microwave Assisted Extraction. *Green Chemistry,* 18: 6482–6492.

Hou, K., Bao, M., Wang, L., Zhang, H., Yang, L., Zhao, H. and Wang, Z. (2019). Aqueous Enzymatic Pretreatment Ionic Liquid–Lithium Salt Based Microwave–Assisted Extraction of Essential Oil and Procyanidins from Pinecones of Pinus Koraiensis. *Journal of Cleaner Production,* 236: 117581.

Jeyaratnam, N., Nour, A. H., Kanthasamy, R., Nour, A. H., Yuvaraj, A. and Akindoyo, J. O. (2016). Essential Oil from Cinnamomum Cassia Bark through Hydrodistillation and Advanced Microwave Assisted Hydrodistillation. *Industrial Crops and Products,* 92: 57–66.

Kusuma, H. S. and Mahfud, M. (2017). Microwave-Assisted Hydrodistillation for Extraction of Essential Oil from Patchouli (Pogostemon Cablin) Leaves. *Periodica Polytechnica Chemical Engineering,* 61: 82–92.

Lee, C. S., Binner, E., Winkworth-Smith, C., John, R., Gomes, R. and Robinson, J. (2016). Enhancing Natural Product Extraction and Mass Transfer Using Selective Microwave Heating. *Chemical Engineering Science,* 149: 97–103.

Li, Y., Fabiano-Tixier, A.-S., Chemat, F., Li, Y., Fabiano-Tixier, A.-S. and Chemat, F. (2014). Essential Oils: From Conventional to Green Extraction. *Essential Oils as Reagents in Green Chemistry,* 9–20. Springer.

Li, Y., Fabiano-Tixier, A. S., Vian, M. A., & Chemat, F. (2013). Solvent-free microwave extraction of bioactive compounds provides a tool for green analytical chemistry. *TrAC Trends in Analytical Chemistry*, 47: 1–11.

Lianfu, Z. and Zelong, L. (2008). Optimization and Comparison of Ultrasound/Microwave Assisted Extraction (Umae) and Ultrasonic Assisted Extraction (Uae) of Lycopene from Tomatoes. *Ultrasonics Sonochemistry,* 15: 731–737.

Liu, B., Fu, J., Zhu, Y. and Chen, P. (2018a). Optimization of Microwave-Assisted Extraction of Essential Oil from Lavender Using Response Surface Methodology. *Journal of oleo Science,* 67: 1327–1337.

Liu, Z., Deng, B., Li, S. and Zou, Z. (2018b). Optimization of Solvent-Free Microwave Assisted Extraction of Essential Oil from Cinnamomum Camphora Leaves. *Industrial Crops and Products,* 124: 353–362.

Lucchesi, M. E., Chemat, F. and Smadja, J. (2004). Solvent-Free Microwave Extraction of Essential Oil from Aromatic Herbs: Comparison with Conventional Hydro-Distillation. *Journal of chromatography A,* 1043: 323–327.

Mandal, V., Mohan, Y. and Hemalatha, S. (2007). Microwave Assisted Extraction – an Innovative and Promising Extraction Tool for Medicinal Plant Research. *Pharmacognosy Reviews,* 1: 7–18.

Martínez-Abad, A., Ramos, M., Hamzaoui, M., Kohnen, S., Jiménez, A. and Garrigós, M. C. (2020). Optimisation of Sequential Microwave-Assisted Extraction of Essential Oil and Pigment from Lemon Peels Waste. *Foods,* 9: 1493.

Mason, T. J., Chemat, F. and Vinatoru, M. (2011). The Extraction of Natural Products Using Ultrasound or Microwaves. *Current Organic Chemistry,* 15: 237–247.

Mazidi, S., Rezaei, K., Golmakani, M., Sharifan, A. and Rezazadeh, S. (2012). Antioxidant Activity of Essential Oil from Black Zira (Bunium Persicumboiss.) Obtained by Microwave-Assisted Hydrodistillation. *Journal of Agricultural Science and Technology, 14*: 1013–1022.

Michel, T., Destandau, E. and Elfakir, C. (2011). Evaluation of a Simple and Promising Method for Extraction of Antioxidants from Sea Buckthorn (Hippophaë Rhamnoides L.) Berries: Pressurised Solvent-Free Microwave Assisted Extraction. *Food Chemistry,* 126: 1380–1386.

Movaliya, S. (2017). Extraction of Essential Oil by Microwaveassisted Extraction: A Review. *Ijariie International Journal,* 3: 2395–4396.

Parizotto, C. A., Dall'Oglio, E. L., de Vasconcelos, L. G., de Sousa, P. T., Taques Filho, E. G. and Kuhnen, C. A. (2019). Measuring Dielectric Properties for Microwave-Assisted Extraction of Essential Oils Using Single-Mode and Multimode Reactors. *RSC Advances,* 9: 5259–5269.

Racoti, A., Buttress, A. J., Binner, E., Dodds, C., Trifan, A. and Calinescu, I. (2017). Microwave Assisted Hydro-Distillation of Essential Oils from Fresh Ginger Root (*Zingiber Officinale* Roscoe). *Journal of Essential Oil Research,* 29: 471–480.

Ramanadhan, B. (2005). *Microwave Extraction of Essential Oils (from Black Pepper and Coriander) at 2.46 Ghz* [Master's thesis, University of Saskatchewan]. HARVEST Institutional Repository.

Routray, W. and Orsat, V. (2012). Microwave-Assisted Extraction of Flavonoids: A Review. *Food and Bioprocess Technology,* 5: 409–424.

Sadeghi, A., Hakimzadeh, V. and Karimifar, B. (2017). Microwave Assisted Extraction of Bioactive Compounds from Food: A Review. *International Journal of Food Science and Nutrition Engineering,* 7: 19–27.

Sahraoui, N., Vian, M. A., Bornard, I., Boutekedjiret, C. and Chemat, F. (2008). Improved Microwave Steam Distillation Apparatus for Isolation of Essential Oils: Comparison with Conventional Steam Distillation. *Journal of Chromatography A,* 1210: 229–233.

Sahraoui, N., Vian, M. A., El Maataoui, M., Boutekedjiret, C. and Chemat, F. (2011). Valorization of citrus by-products using Microwave Steam Distillation (MSD). *Innovative Food Science & Emerging Technologies,* 12(2): 163–170.

Stashenko, E. E., Jaramillo, B. E. and Martınez, J. R. (2004). Analysis of Volatile Secondary Metabolites from Colombian Xylopia Aromatica (Lamarck) by Different Extraction and Headspace Methods and Gas Chromatography. *Journal of Chromatography A,* 1025: 105–113.

Tang, J. and Resurreccion, Jr, F. (2009). Electromagnetic Basis of Microwave Heating. *Development of Packaging and Products for Use in Microwave Ovens*, 3–38e. Elsevier, Woodhead Publishing.

Tatke, P. and Jaiswal, Y. (2011). An Overview of Microwave Assisted Extraction and Its Applications in Herbal Drug Research. *Research Journal of Medicinal Plant,* 5: 21–31.

Uysal, B., Sozmen, F. and Buyuktas, B. S. (2010). Solvent–Free Microwave Extraction of Essential Oils from Laurus Nobilis and Melissa Officinalis: Comparison with Conventional Hydro-Distillation and Ultrasound Extraction. *Natural Product Communications,* 5: 1934578X1000500127.

Vian, M. A., Fernandez, X., Visinoni, F. and Chemat, F. (2008). Microwave Hydrodiffusion and Gravity, a New Technique for Extraction of Essential Oils. *Journal of Chromatography A,* 1190: 14–17.

Wang, Y., Li, R., Jiang, Z.-T., Tan, J., Tang, S.-H., Li, T.-T., Liang, L.-L., He, H.-J., Liu, Y.-M. and Li, J.-T. (2018). Green and Solvent-Free Simultaneous Ultrasonic-Microwave Assisted Extraction of Essential Oil from White and Black Peppers. *Industrial Crops and Products,* 114: 164–172.

Yang, L. M., Han, L. L. and Yang, Z. (2012). Microwave-Assisted Extraction of Garlic Essential Oil from Garlic. *Applied Mechanics and Materials,* 117: 1022–1026.

7 Microwave-Assisted Extraction

A Green Approach to Obtaining Essential Oils

Rupesh Kaushik, Harjinder Kaur, Ab Waheed Wani, Ajay Saroha, Sanjeev Kumar and Sapna Jarial

7.1 INTRODUCTION

Over the past ten years, the food processing sector has observed an advancement in the creation of more eco-friendly techniques for the recuperation of active ingredients due to the scientific, technical, and economic obstacles that accompany traditional methods of extraction, like high-energy costs of energy, residuals solvent impurities, and thermal degradation (Alexandre et al., 2017; Chemat et al., 2020; Pagano et al., 2021; Zhu et al., 2016). Examples of such innovative methods for selecting, separating, and getting different foodstuffs include enzyme-assisted extraction (EAE), high-voltage electrical discharges, microwave-assisted extraction (MAE), ultrasound-assisted extraction (UAE), high-pressure processing, and supercritical fluid extraction (SFE), as well as their combinations. MAE is the technique for recovering food ingredients that have drawn the most interest because of its superior extraction efficiency, sustainability, low cost, shorter recuperation time, and installation flexibility on small and large scales. In the first demonstration of recuperation, a study was done to determine the effects of microwave irradiation on organic compounds (Ganzler et al., 1986). Since then, several projects for improvement have been created and implemented to boost MAE's operational effectiveness and sustainability. A typical MAE technique comprises material penetration by microwave radiation, its interaction with polar molecules, and substantial heating caused by ionic conduction or dipole rotation, which result in a high yield.

Nevertheless, to ensure continuity, certain limitations include eliminating the solvent from the substance once the process is finished, and its restriction to polar extractants ought to be removed (Shen et al., 2021). Research findings suggest that existing closed systems and open MAE set-ups can be enhanced by incorporating various add-ons or integrating them with complementary technologies. These

DOI: 10.1201/9781032634180-7

modifications aim to improve the efficiency of the extraction process further. Historically, water or an organic solvent was used as a solvent in MAE systems, but these days there are many types of hydrotropic fluids, dual-phase solutions, and miscellars (Gao et al., 2020; Shen et al., 2021; Baite et al., 2021; Kumar et al., 2021; Zhu et al., 2022; Saha et al., 2022). These methods offer an environmentally friendly implementation strategy by using fewer harmful solvents, producing less waste, emitting less carbon dioxide (CO_2) into the atmosphere, and using less energy. Additionally, some hybrid modes, like Microwave hydrodiffusion and gravity (MHG), guarantee solvent recycling, reuse extraction by-products in the following processes, and increase extraction efficiency (Panzella et al., 2020). Given the numerous reviews of MAE systems published in the past (Mandal &Tandey, 2016), it is obvious how much emphasis has been placed on the extraction set-up. The development of MAE systems has lengthily progressed from simple methods, or their temporary alterations, to the use of more environmentally friendly based on solvents. MAE systems and their combination of additional cutting-edge or novel techniques significantly increase the entire procedure efficiency in a previously unheard of, based on a detailed examination and results from recent experimental studies. Although researchers pointed out that a few of these innovations engaged the use of environmentally conscious analytical solvents, it is critical to modify the data that are currently accessible regarding these entirely recently invented combination techniques for MAE systems and environmentally friendly solvents to support their viability and the requirement of process validation and optimisation in a step-by-step manner. Because of this, this chapter presents a collection of newly released research that focuses on cutting-edge developments in MAE systems and approaches. Different extractants, combination strategies, and subsequent applications throughout the extraction of the intended compounds are specifically addressed. The installation and setting up of the equipment are also addressed, along with the principles and methods. It also details how MAE technology has evolved and where it is headed. This chapter emphasises their popularity, effectiveness, and movement from benchtop to industrial application.

7.2 ESSENTIALS OF MAES

7.2.1 Principles and Workings of MAE

The way two simultaneous processes convey energy by electrical field justifies the employment of microwaves for energy extraction. Dipoles interact with polar components in the first instance, resulting in the dipoles realigning, reorienting themselves with the applied field, and producing heat (Kumar et al., 2021; Chaves et al., 2020). Second, after using electromagnetic radiation, charged ions in the solvent move due to ionic conduction. The solution's internal resistance leads to friction, which generates heat. However, the food matrix's ability to create heat and absorb microwave radiation is impacted by its dielectric loss. When microwave energy interacts with small quantities of fluids within the cellular matrix, it induces evaporation. The cell wall breaks down under intense pressure, releasing the active ingredients. It is possible to generate higher throughputs by carefully selecting working parameters, such as temperature, which promotes quicker solvent penetration into the cell matrix. Like polar molecules, ionic mixtures and polar molecules readily absorb microwave radiation due to an ongoing dipole moment that interacts with them. Non-polar solvents, like hexane, do not spontaneously heat up when interacting with microwaves. The solid-to-liquid ratio, the duration of the extraction, the microwave's strength, the nature of the materials being extracted, and swirling are additional variables that determine how well MAE systems function. You may get more information on the influence of these settings elsewhere (Chaves et al., 2020).

7.2.2 MAE System Configuration and Instrumentation

A standard simplified MAE system typically includes components such as a cavity, a waveguide, an isolator, a mode stirrer, and a magnetron. The mode stirrer distributes the energy in various ways after the magnetron generates microwave radiation that travels through the waveguide and into the cavity. Until the substances matrix entirely consumes the energy, it is contained in the cavity. The isolator allows energy to pass from the magnetron to the hole but not either way around, preventing energy that is reflected from accessing the magnetron and lowering power output. Depending on whether they operate at pressures above or below atmospheric pressure, MAE devices are usually classified into two groups: "closed systems" (pressurised MAE) or "open systems" (focused MAE). The vessel's high pressure, which has been meticulously maintained, enables quick and effective evacuation. The MAE extraction procedure is sped up by a concurrent rise in temperature and pressure caused by higher solvent microwave energy absorption. Closed systems also have a confined analyte output and vulnerability to temperature deterioration, even though they have an elevated extraction efficiency.

On the other hand, open microwave-aided extraction systems provide benefits including fewer safety issues, little degradation of sensitivity to heat chemicals, higher efficiency, and the capability to add solvent at any moment during extraction. Additionally, the system includes a return apparatus that condenses the vaporized solvent, operates under ambient pressure, and functions in either mono- or multi-mode. Although open MAE systems have an acceptable degree of efficiency, as was already said, several drawbacks have led different research teams to develop MAE systems having greater degrees of efficacy (Fierascu et al., 2023; More et al., 2022; Gao et al., 2020).

7.2.3 MAE Systems Intermediary Alterations

Since its creation, the basic MAE system has undergone several adjustments for better average extracting performance. These improvements were made by either altering the fundamental design or adding a new operational scenario or system.

For instance, in a study (Louati et al., 2020), an offline MAE was converted to a closed MAE. The process was carried out in a microwave with a container with elevated pressure to inject solvent into the system. The sample is treated with a microwave after immersing it in the solvent. After extraction, a portion of the extracts are aspirated by a different pump employing a membrane that traps analytes while releasing the remaining portions as waste. Following an acetonised steam rinse, the analytes are measured quantitatively using a light-scattering detection device. Since inert gases like nitrogen can inhibit the reactions of active components with oxygen, a closed MAE system was modified by Bagade et al. (2021) to operate under nitrogen protection to increase process efficiency and decrease losses. The microwave-aided extraction underneath nitrogen gas (NPMAE) approach was utilised to improve the synthesis of vitamin C from specific fruits and vegetables. A quantitative estimate was then made by employing high-performance liquid chromatography.

In contrast to the techniques described above, the MAE apparatus, also known as dynamic MAE, can be connected to an analytical step to permit mechanisation and parallel operation. Compared to batch MAE, this continuous technique eliminates the need for an extraction cycle, minimises contaminant or analyte loss risk, and slows compound heat degradation. According to Chaves et al. (2020), flavonoids from *H. epimedii* and safflower yellow from Flos Carthami have been swiftly extracted using dynamic MAE.

7.3 CHEMISTRY, EXTRACTION, AND USES OF ESSENTIAL OILS AND AROMAS

Essential oils are made using physical extraction methods from raw plant sources. These methods include steam, steam/water, or dry distillation (pyrolysis) of organic compounds, and cold-pressed extraction for peelings of citrus oils. Distilling is used to separate the essential oil from the aqueous phase physically. The ability of volatile compounds to disintegrate in fatty oils and fats is called the "essential oils" concept. "Oil" refers to the plant's hydrophobic and viscous properties. In contrast, "essential" refers to the plant's inherent essence and characteristic aroma. The traditional method for obtaining essential oils from plant matter is distillation, which involves exposing fragrant plants to steam or boiling water. Recovering essential oils using distillation is simpler, predicated on the notion that the ambient pressure equals the total vapour pressure at boiling point. As a result, crucial oil constituents can evaporate at about the same temperature as water, typically between 200°C and 300°C. In a chilly, constricted tube, the rising steam that contains the essential oil condenses and is collected. Plants usually produce 0.005% to 10% of their total oil. These oils come from various aromatic plants and spice components and are intensely strong, aromatic, oily fluids (Mandal &Tandey, 2016).

Several publications have offered information on the composition of the many essential oils comprising over one hundred distinct elements. The two fundamental groups into which essential oils can be separated are hydrocarbons, which include terpenes like monoterpenes, diterpenes, sesquiterpenes, and oxygenated compounds, which include ester compounds, ketones, aldehydes, alcohols, phenols, oxides, acidic substances, and lactones. Sulphur and nitrogen-containing substances can also exist. Terpenes are a group of naturally occurring substances having the chemical formula $(C_5H_8)_n$ and a structure based on isoprene (2-methyl butadiene) units. This term may also be used to describe the terpenoids, which are oxygenated derivatives of these compounds. Two isoprene units make up monoterpenes, and terpenes are frequently classed based on how many isoprene units they are biogenetically formed from (Shen et al., 2021; More et al., 2022). Terpenes, common in essential oils and abundant in nature, play important functions in the flavour and scent industries. Three isoprene units comprise the sesquiterpenes and are present in various living things, notably higher plants. Four isoprene units can combine to form 20-carbon chemical molecules known as diterpenes. They are present in nearly all plant groups, covering 20 or more primary structural kinds. Plants produce aromatic chemicals as end products or by-products and store them in specific plant organs (Table 7.1). About 300 known essential oils are economically valuable and commonly utilised in the food, agriculture, beauty products, pharmaceuticals, and perfumery industries (Table 7.2). Numerous of these oils have long been known to have antioxidant and antimicrobial effects that are insecticidal, antiviral, toxic, anti-parasitic, and anti-mycotic (Yousuf et al., 2021). These traits relate to the roles that these compounds play in crops. For instance, whereas monoterpenes and sesquiterpenes are relatively non-toxic to mammals, they are toxic to insects and have anti-herbivore effects. According to Yousuf et al. (2021), essential oils can also be a source of compounds used in tastes and perfumes.

TABLE 7.1
Essential oils produced by plant organs.

Organ of the plant	Example
Berries	Black pepper, juniperus
Seeds	Fenugreek, dill, coriander, caraway
Arils	Mace of nutmeg
Roots	Angelica
Pistil	Saffron
Barks	Cinnamon, tea tree
Latex from rhizome	Asafoetida
Kernel	Nutmeg
Buds	Clove
Leaf	Basil, mint, sage, geranium
Bulbs	Leek, onion, garlic
Flower	Rose, neroli
Rhizome	Ginger, iris

TABLE 7.2
Vital flavour components in essential oils.

Essential oil	Flavour components
Spearmint	(L)-Carvone
Parsley	Apiol
Caraway	Carvone, derivatives of carvone
Anise	(E)-Anethole, methyl chavicol
Peppermint	Menthone, menthfuran, Menthol
Allspice	Eugenol, β–caryophyllene
Black pepper	Piperine, δ3-carene
Cinnamon	Cinnamaldehyde
Mustard	Allyl isothiocyanate
Coriander	Linalool

7.4 TECHNIQUES FOR EXTRACTING ESSENTIAL OILS AND AROMA

7.4.1 Microwave Compressed Air Distillation

The first microwave technique to be utilised for this purpose, the microwave-based extraction method for obtaining essential oils from fragrant spices and herbs, was proposed by Craveiro et al. in 1989. The three main components of the system for extraction are a microwave, a refrigeration system, and a compressor for air that pushes air into the vessel containing the plant matrix (Figure 7.1). In contrast to the steam extraction technique, which relies on the thermal condensation principle, compressed air is used to extract the essential oil. In this process, compressed air is continually fed into a reactor that contains a plant substrate immersed in water. After getting saturated with volatile chemicals, the steam produced by the microwave-based extraction technology is subsequently transferred to a vessel for recovery that is submerged in a system of cooling the exterior of the microwave oven. Both water and aromatic molecules soon condense and are recovered in ratios similar to conventional techniques. A similar invention with the name "Microwave equipment and technique for harvesting aromatic oils, their essence, and dyes with gas flow" was patented by Lee in 2010. To cool the gas leaving the extraction cartridges, the condenser utilised in this type of system runs between −20°C and −15°C. This extraction method is eco-friendly because it does not employ artificial chemicals or organic solvents (Stashenko et al., 2004a).

7.4.2 Hydro-distillation by Microwave

Stashenko et al. (2004b) created the microwave hydro-distillation (MWHD) method as shown in Figure 7.2. It is based on the conventional hydro-distillation

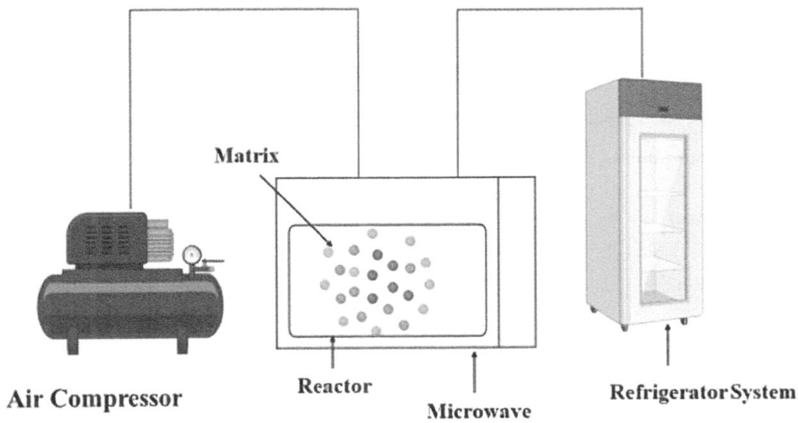

FIGURE 7.1 Microwave Compressed Air Distillation.

Source: Photograph by the author.

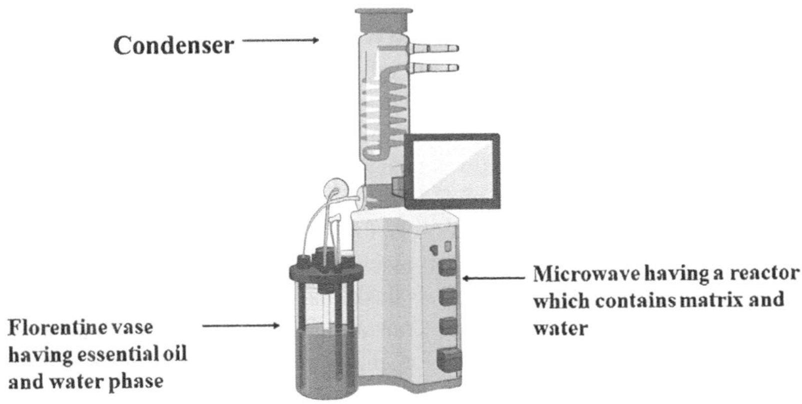

FIGURE 7.2 Microwave hydro-distillation.

Source: Photograph by the author.

theory and involves placing a piece of the hydro-distillation apparatus inside a microwave oven. This technique involves mixing the plant matrix with water and placing it in a reactor within the microwave oven. One can find the cooling system and essence recovery component outside the oven. A standard microwave oven is transformed into a hydro-distillation system, with a side orifice linking a glass condenser outside the oven to a circular flask that contains water and the matrix (*Xylopia aromatica*) (100 g). The water vigorously reflux boils when the oven is run at 800 W for 30 minutes. The process of obtaining essential oils from various aromatic herbs and spices using the MWHD method has been the

TABLE 7.3
Utilising MWHD to extract essential oils.

Botanical species	Extraction operation conditions	References
Zataria multiflora Boiss.	60 g, 1,200 ml water, $R = 3.7\%$, 990 W, P(atm), $T = 120$ min,	Tasa et al. (2020)
Satureja montana	60 g, 1,200 ml water, $R = 0.7\%$, 660 W, P(atm), $T = 90$ min,	Radivojac et al. (2021)
Xylopia aromatica (Lamarck)	100 g, 2,000 ml water, 800 W, P(atm), $T = 30$ min	Addo et al. (2021)
Anethum graveolens L.	100 g, 1,000 ml water, $R = 2\%$ 500 W, P(atm), $T = 60$ min,	Ghazanfari et al. (2020)

subject of substantial investigation. For instance, during MWHD, as shown in Table 7.3, *Thymus vulgaris* L. was subjected to the following conditions: 60 g of plant matrix, 2,000 ml of water, 990 W of microwave power, and a yield of 2.52% essential oil were used (Bora et al., 2020). *Zataria multiflora* Boiss is an example of a plant subjected to extraction using MWHD. In this process, the plant material (60 g) was combined with water (1,200 ml) as the solvent, and the extraction was conducted at atmospheric pressure with microwave heating at 990 W for 120 minutes. This extraction yielded 3.7% essential oil (Stephane & Jules, 2020).

Similarly, *Satureja montana* was also subjected to extraction using MWHD. The plant material (60 g) was mixed with water (1,200 ml) as the solvent, and microwave heating was applied at 660 W for 90 minutes under ambient pressure. The extraction process yielded 0.7% essential oil (Guo et al., 2021). The plant matrix and water were placed in a glass flask, and an insulated microwave coaxial antenna was inserted into the flask; Constantino et al. (2022) used an improved version of MWHD. This technique for *in-situ* microwave heating is considered secure and adaptable, providing benefits including time and energy savings, making it appropriate for industrial uses. Microwave steam distillation (MSD), as shown in Figure 7.3, a different invention, was created by Sahraoui et al. (2008). With the addition of microwave radiation focused just on the extraction reactor, MSD uses the traditional steam distillation (SD) principle. The cooling system and the element in charge of essence recovery are situated outside the oven, just like MWHD. Lavender (*L. angustifolia* Mill.) blossom essential oil was extracted using this technique.

7.4.3 Microwave Extraction Without Solvent

SFME, or solvent-free MAE, was patented by Chemat et al. in 2004. It represents one of the most recent advancements in oil extraction using microwave-assisted techniques, as displayed in Figure 7.4. SFME operates at air pressure and without the need for water or solvents. The reactor, where the matrix of plants to be

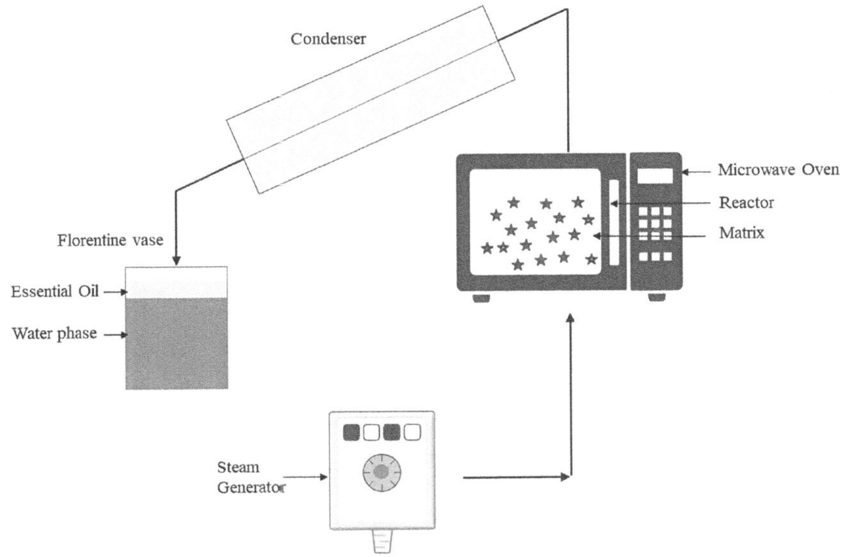

FIGURE 7.3 Microwave steam distillation.

Source: Photograph by the author.

addressed is set up, the microwave appliance, the chilling system, and the essence vessel used to bring the oil are the four critical components of the SFME process. This process uses a microwave-assisted dry distillation technique founded on a relatively straightforward concept. It requires putting the fresh plant matrix into a microwave reactor with no organic fluids or solvents. The essential oil glands within the raw material are ruptured when the water is boiled. This action releases the essential oil, transported by the steam the water creates in the matrix. A cooling system has been installed outside the microwave oven to ensure water and essential oil are continuously condensed into the distillate. The matrix's moisture content is maintained appropriately by returning surplus water to the flask. SFME is conducted using a Milestone "DryDist" microwave laboratory oven, which has a maximum power output of 1,000 W. This unusual oven operates at 2,450 MHz and is capable of multi-mode microwave operations. Without water or solvents, the experiment is carried out at atmospheric pressure using 250 g of the plant matrix for 30 minutes at 500 W. The essential oil was extracted at a temperature of 100°C until no more could be extracted. After being collected, the essential oil was dried with anhydrous sodium sulphate and stored at 4°C. Then, the essential oils were examined using gas chromatography and mass spectrometry. Basil, crispate mint, and thyme were reported to have yields of essential oils by SFME at rates of 0.029%, 0.095%, and 0.160%, respectively (Chemat et al., 2020). Citrus fruits, spices, aromatic herbs, and other dried and fresh plant types have all been successfully extracted using this technique (Rao et al., 2021; Lefebvre et al., 2021; Rasekh et al., 2021; Hossain et al., 2021; Gholamian et al., 2021; Guo et al., 2021).

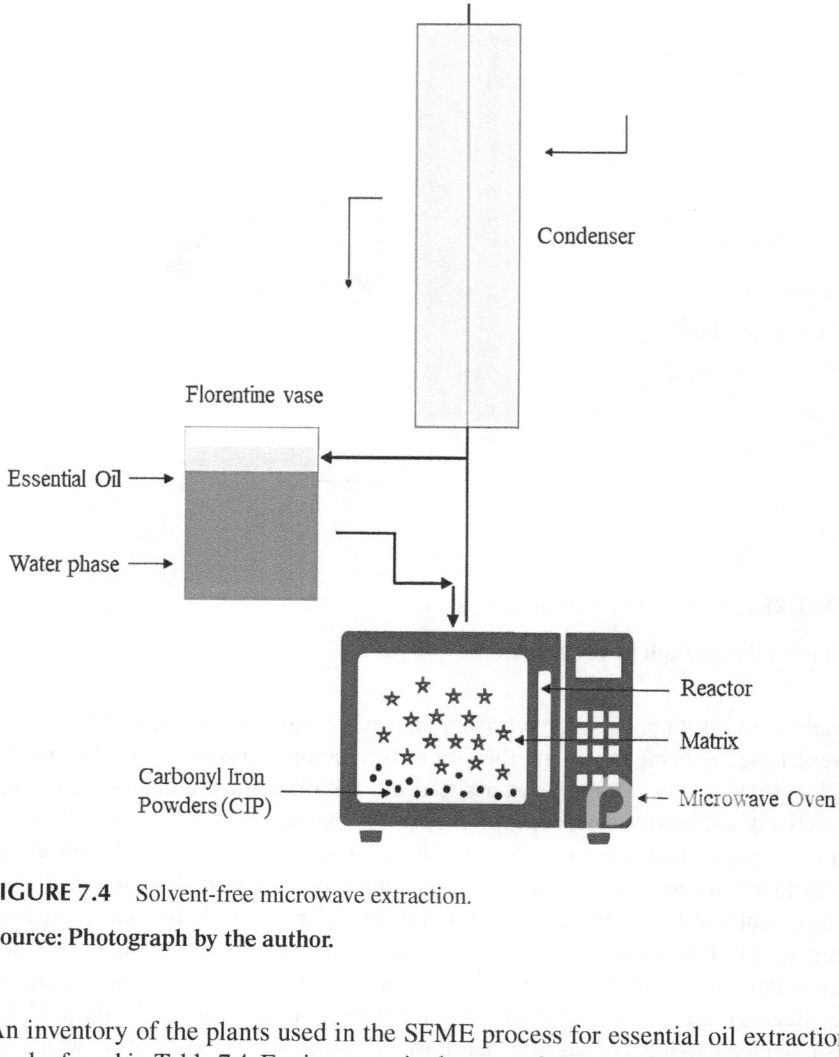

FIGURE 7.4 Solvent-free microwave extraction.

Source: Photograph by the author.

An inventory of the plants used in the SFME process for essential oil extraction can be found in Table 7.4. For instance, in the case of *Ocimum basilicum* L., 250 g of the plant material was extracted at 500 W for 30 minutes at atmospheric pressure, yielding 0.029% essential oil (Rao et al. 2021). The spherical carbonyl iron powder (CIP) particles can collect a portion of the radiation from the microwave and release it as heat, allowing the matrix to be warmed by simple conduction without additional energy sources. Graphite particles, activated carbon powders, and an ionic liquid (IL) known as 1-hexyl-3-methylimidazolium hexafluorophosphate are some of the materials that can absorb microwave radiation (Mukherjee et al. 2022; Yu et al. 2021). The expanded SFME technique has been successfully applied to various dry and fresh plants, including spices, as shown in Table 7.5, citrus fruits, and fragrant herbs (Mukherjee et al., 2022; Yu et al., 2021). For instance, in the case of *Cuminum cyminum* L., 100 g of the substrate and 20 g of CIP were heated at 85 W for 30 minutes while under the atmosphere's pressure

TABLE 7.4
List of plants subjected to SFME to extract essential oils.

Name of species	Operational conditions for extraction	References
Calamintha nepeta L. Savi	60 g, 250 W, *R* = 0.38%, P(atm), *T* = 40 min,	Mahanta et al. (2021)
Elletaria cardamomum L.	100 g immersed in water (67% of humidity), 390 W, *R* = 2.70%, P(atm), *T* = 75 min,	Lefebvre et al. (2021)
Rosmarinus officinalis L.	250 g, 500 W, *R* = 0.39%, P(atm), *T* = 40 min,	Skendi et al. (2022)
Laurus nobilis L.	140 g immersed in water, 85 W, P(atm), *R* = 0.42%, *T* = 50 min,	Ordoudi et al. (2022)
Ocimum basilicum L.	250 g, 500 W, *R* = 0.029%, P(atm), *T* = 30 min,	Rao et al. (2021)
Bupleurum fruticosum L.	500 W, *R* = 0.21%, P(atm), *T* = 30 min,	Zhao et al. (2023)
Melissa officinialis L.	280 g immersed in water, 85 W, *R* = 0.15%, P(atm), *T* = 50 min,	Ordoudi et al. (2022)
Mentha crispa L.	250 g, 500 W, *R* = 0.095%, P(atm), *T* = 30 min,	Rao et al. (2021)
Origanum vulgare L.	150 g immersed in water during 1 h, 622 W, *R* = 0.054 ml/g, P(atm), *T* = 35 min,	Gholamian et al. (2021)
Nigella sativa L.	150 g, 850 W, *R* = 0.20%, P(atm), *T* = 10 min,	Hossain et al. (2021)
Rosmarinus officinalis L.	200 g, 200 W, *R* = 0.57%, P(atm), *T* = 30 min,	Rasekh et al. (2021)
Origanum glandulosum Desf.	25 g immersed in water for a full hour, 850 W, *R* = 3.3%, P(atm), *T* = 20 min,	Guo et al. (2021)

(Mukherjee et al., 2022). *Citrus reticulata* Blanco was extracted using the same operating conditions of 100 g of substrate and 20 g of CIP at 85 W for 30 minutes under ambient pressure (Mukherjee et al., 2022). In the case of *Mentha haplocalyx* Briq., the same extraction techniques were applied, utilising 100 g of the plant material and 20 g of CIP, heated at 85 W for 30 minutes under the atmosphere's pressure (Mukherjee et al., 2022).

7.4.4 Microwave Hydrodiffusion and Gravity

The patented technique of MHG extraction was developed by Chemat et al. (2008). Only the hydro-diffusion-based, ambient-pressure extraction of aromatic oils from a range of matrices is the intended use of this innovative method. The MHG system (Figure 7.5) is made up of four fundamental components: a reactor

TABLE 7.5
List of plants subjected to an improved SFME for extracting essential oils.

Botanical species	Operational conditions for extraction	References
Zanthoxylum bungeanum Maxim.	P(atm), *T* = 30 min, 100 g + 20 g CIP, 85 W	Zia et al. (2022)
Cuminum cyminum L.	P(atm), *T* = 15 min, 20 g + 15 ml IL, 440 W	Yu et al. (2021)
Mentha haplocalyx Briq.	P(atm), *T* = 30 min, 100 g + 20 g CIP, 85 W	Mukherjee et al. (2022)
Illicium verum Hook. f.	P(atm), *T* = 15 min, 20 g + 15 ml IL, 440 W	Yu et al. (2021)
Citrus reticulata Blanco	P(atm), *T* = 30 min, 100 g + 20 g CIP, 85 W	Erukainure et al. (2021)

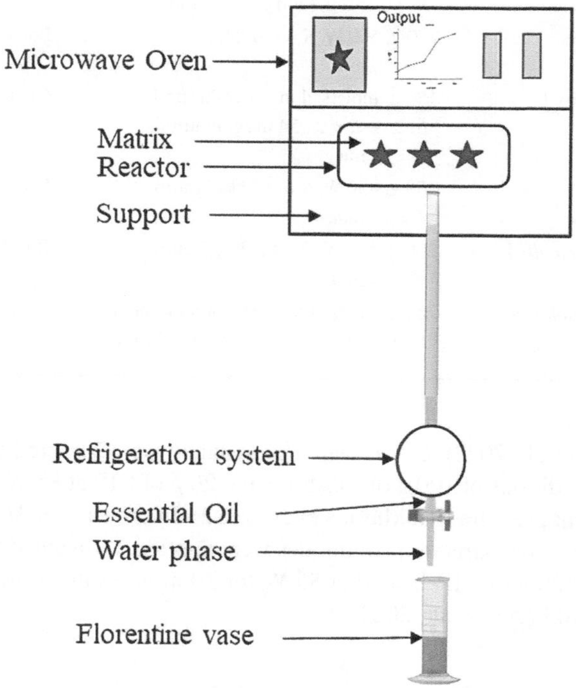

FIGURE 7.5 Hydrodiffusion microwave and gravity.

Source: Photograph by the author.

to hold the prepared matrix, a microwave, a cooling system, and a container to store the extracted essential oil. The matrix is placed in a microwave reaction chamber without water or chemical solvents. The fluid in the matrix rises when microwaves are employed, which breaks the cells that carry the essential oils. About the hydro-diffusion phenomenon, volatile oils and intracellular water from the matrix of plants can be liberated and transmitted to the surrounding environment. A cooling system has been placed underneath the microwave chamber to reduce precipitation. The MHG extraction method is carried out in the Milestone NEOS-GR microwave laboratory oven with a 900 W maximum output. This oven functions as multiple modes of microwave reactor with an operating frequency of 2.45 GHz. An outside infrared sensor keeps track of the temperature. Without utilising any water or solvents, the experiment is run at the atmosphere's pressure with a 500 g matrix and an extraction time of 20 minutes at 500 W.

The resulting essential oil is extracted, dried with a sodium sulphate solution, and stored at 4°C. According to Sharma et al. (2021), the MHG extraction method yields 0.6% for spearmint and 0.95% for pennyroyal while offering advantages, including lower costs and time and energy savings. Citrus species and aromatic plants had their essential oils successfully extracted using the MHG method (Sharma et al., 2021; Yousuf et al., 2021). Table 7.6 lists the plants used in the MHG process for extracting essential oils. A yield of 0.95% essential oil was obtained when *Mentha pulegium* L. was extracted after heating 500 g of the substrate at 500 W for 20 minutes under air pressure (Sharma et al., 2021). Similar results were obtained when 500 g of *Citrus limon* L. material was treated at 500 W for 15 minutes (Yousuf et al., 2021), yielding 0.7% and 1.6% aromatic oil, under atmospheric pressure. In a different experiment, 500 g of the plant material *Rosmarinus officinalis* L. (Sharma et al., 2021) was heated to 500 W for 15 minutes, producing 0.33% of the essential oil.

TABLE 7.6
List of plants subjected to an MHG essential oil extraction.

Name of species	Operational conditions of extraction	References
Citrus limon L.	500 g, P(atm), R = 0.7%, 500 W, T = 15 min,	Sharma et al. (2021)
Menthe pulegium L.	500 g, P(atm), R = 0.95%, 500 W, T = 20 min,	Sharma et al. (2021)
Citrus sinensis L.	500 g, P(atm), R = 1%, 500 W, T = 15 min,	Yousuf et al. (2021)
Citrus aurantifolia (Chrism.) Swingle	500 g, P(atm), R = 0.8%, 500 W, T = 15 min,	Yousuf et al. (2021)
Citrus limon L.	500 g, P(atm), R = 1.6%500 W, T = 15 min	Yousuf et al. (2021)
Menthe spicata L.	500 g, P(atm), R = 0.6% 500 W, T = 20 min,	Sharma et al. (2021)
Citrus sinensis L.	500 g, P(atm), R = 1.2% 500 W, T = 15 min,	Yousuf et al. (2021)
Citrus sinensis L.	500 g, P(atm), R = 0.9%, 500 W, T = 15 min,	Yousuf et al. (2021)

7.4.5 Microwave-Assisted Distillation

Rasekh et al. (2021) utilised microwave-assisted distillation as a contemporary approach for extracting essential oils from plants, which is now widely used in modern extraction methods. This procedure accelerates the distillation process through microwave energy, leading to a quicker and more effective extraction. The application of microwave radiation to heat the plant material and enhance the extraction of essential oils is the basis of the microwave-assisted distillation process. Preparing the plant material is the first stage in the microwave-assisted distillation process. The plant material has been expanded to increase the surface area and enhance the release of essential oils, such as leaves, flowers, or stems, which are often dried and crushed. After that, the processed plant material is put in an appropriate distillation vessel, like a round-bottom flask or a microwave-safe container. Then water is introduced to the vessel, combining with the plant matter. After that, the distillation vessel is placed in a microwave, which supplies the extraction process with the necessary microwave energy.

The combination is heated by microwave energy, which causes the water to evaporate (Stephane & Jules, 2020). The components from the essential oils are carried by the water when it vaporises. The mixture is then typically passed through a condenser as it is condensed in a different collection jar. The condenser cools the vapour, which causes it to return to liquid form. Water and essential oil comprise the condensed liquid, with the essential oil settling to the top because of its lower density. The liquid is poured into a separate funnel or flask to separate the essential oil from the water. The essential oil will naturally separate and create a unique layer on top of the water because oil and water do not mix. The essential oil can be recovered when the water has been removed from the bottom. To preserve its quality and shield it from light, the essential oil is gathered and placed in an appropriate container, such as an amber glass bottle. The oil may be subjected to additional examination, purification or utilised immediately for several purposes, including aromatherapy, fragrance, and flavouring. Compared to conventional distillation techniques, microwave-assisted distillation has several benefits. When microwave energy is used, the extraction process can frequently be completed in minutes or hours rather than days. Additionally, it uses less water and plant material, making it a more environmentally responsible choice. Additionally, because the extraction process takes less time, heat-sensitive molecules are less likely to degrade, producing essential oils of greater quality (Stephane & Jules, 2020).

7.4.6 Pressurised MAE

A cutting-edge extraction method called pressurised MAE (PMAE) combines high-pressure and microwave energy to extract chemicals from various materials effectively. The sample material is placed in a sealed vessel, pressure is increased, and microwave irradiation is used to speed up the extraction process. In comparison to conventional extraction techniques, it has many benefits. Pressure and microwaves greatly shorten the extraction period, saving time and energy. Even at greater temperatures, pressure is applied to maintain the solvent's liquid state, increasing solubility and extraction effectiveness. Microwave technology helps break down cell walls, facilitating the release of target compounds and improving mass transfer, leading to enhanced extraction yields.

The process often includes the following steps: the plant material, such as dried herbs or botanicals, is loaded into a sealed extraction vessel. The plant material extraction vessel is then filled with a suitable solvent, such as water or organic solvents. The dielectric constant must be high for a solvent to absorb microwave energy successfully. The extraction vessel is pressurised to a predetermined level using an inert gas, like nitrogen. Pressurisation aids in raising the solvent's boiling point, enabling extraction at hotter temperatures. The extraction vessel is exposed to microwave energy, which causes the solvent to heat quickly through ionic conduction and dipolar rotation. Due to internal pressure created by the plant cells' fast heating, essential oils are more easily released: the heated, pressurised solvent permeates the plant material and dissolves the chemicals included in essential oils. The increased pressure's ability to keep the solvent in a liquid state at higher temperatures enhances the target chemicals' solubility and diffusion. The extraction process is stopped by cooling the extraction vessel and turning off the microwave energy when the desired extraction time has passed. The pressure inside the vessel is then gradually released to complete the depressurisation process. The remaining plant matter and solvent are removed from the essential oil extract. Several processes can be utilised to separate the essential oil, including filtration and centrifugation.

7.4.7 Subcritical and SFE with Microwave Assistance

The supercritical state refers to a unique phase of a substance achieved when it is subjected to pressure and temperature conditions surpassing its critical point (Georgiopoulou et al., 2022). Supercritical fluids are appropriate for isolating food components, particularly oil and fats, because of their increased viscosity, solvation capacity, and surface tension. SFE also operates at low temperatures and does not leave any solvent behind. One drawback of SFE is the non-polarity of CO_2, the most commonly used supercritical fluid. However, this limitation can be overcome by incorporating other modifiers.

In contrast, subcritical water extraction (SWE) offers an alternative method that keeps liquids in their liquid state by using water as the extractant and heating it to temperatures between 100°C and 374°C under high pressure (Carneiro & Dias, 2021). It has attracted a lot of attention over time as a reliable solvent for both polar and non-polar chemical extractions. It is often referred to as low-polarity water or pressured water. Modern MAE enhancements that increase extraction efficiency include subcritical or SFE. Using *Chlorella vulgaris* as an example, Georgiopoulou et al. (2022) showed that pre-treating it with microwaves before supercritical CO_2 extraction boosted the generation of fatty acids from the sample matrix. Microwave-assisted SWE (MW-SWE) in a single stage was utilised by Carneiro and Dias (2021) to simultaneously determine inorganic and metallic components of selected milieu samples using subcritical water for MAE extraction. The extraction solvent was ethanol–water modified with 1% HNO_3, and the process parameters were a 10-minute microwave treatment, 280°C temperature, 90 bar pressure, and these conditions. According to the results, the method may increase productivity, decrease operating times, and reduce associated operational costs. In addition to recovering metallic compounds, MW-SWE can also retrieve various polar and non-polar molecules, including polyphenols. The biological activity of polyphenols is demonstrated by their ability to reduce conditions caused by excess oxygen radicals. Sarker et al. (2021) discovered that applying a microwave pretreatment at a temperature of 200°C before SWE effectively enhanced the extraction of polyphenols from defatted rice bran by over 55% in just 10 minutes while having greater antioxidant activity than untreated samples (Zhang et al., 2020).

7.4.8 Enzymatic Extraction with Microwave Assistance

EAE is a viable alternative to classical extraction due to its unusual advantages, including friendly to the environment, decreased solvent consumption, enhanced extraction efficiency, and simplified operation (Gong et al., 2020). Alara et al. (2021) contend that the essential mechanisms for efficient extraction depend on the inherent properties of enzymes, such as their high selectivity, acute accuracy, and aptitude to catalyse particular reactions. Ingredients from canola, sunflower, olive, sesame, and other plants have been recovered using EAE. Alara et al. (2021) claim that they are further employed in juice processing to compromise the strength of the cell wall and enhance the release of phenolic components, extending product shelf life and enhancing general quality. Some of the major disadvantages of EAE include its expensive nature, difficulties in scaling up the process for commercial purposes, and the limited effectiveness of certain enzymes in breaking down cells, resulting in lower extraction yields. One effective strategy for addressing these issues is to use microwave irradiation during EAE. According to studies (Kumar et al., 2021), the method known as microwave-assisted enzymatic extraction (MAEE) has a bright future because, as heat is constantly provided, the power supplied by the non-ionising radiation of microwaves causes the disintegration of the cell matrix and allows for the rotation of dipoles. Influencing factors such as the solid-to-liquid ratio, process time, enzyme structure, and food substance dimension should be considered to achieve desired performance. There have been several studies published about MAEE. The most current study by Nguyen et al., finished in 2017, aimed to produce fish protein hydrolysates (FPH) from rainbow trout by employing alcalase as the hydrolytic enzyme. The study explored various enzyme substrate ratios (E:S) and conducted hydrolysis in a microwave system operating at a power level of 1200 W. The objective was to investigate the effects of these variables on the hydrolysis process and the characteristics of the resulting FPH. According to Nguyen et al. (2017), it was hypothesised that microwave heating could enhance solubility and induce structural changes in proteins, leading to the release of active amino acids.

However, Qiu et al. (2022) extracted polysaccharide hydrolysates from *Pyropia yezoensis* using a simplified digestive system, various microwave energies, and hydrolytic conditions under carbohydrates. The algal polysaccharide was synthesised with high efficiency in 2 hours using a 400 W microwave and subsequent ultrafiltration at a 10:1 substrate-to-enzyme ratio.

7.5 SCALABILITY ISSUES

7.5.1 Cost, Energy, and Environmental Issues

The suggested microwave extraction techniques for essential oils have significant energy and time savings, resulting in lower costs. The energy consumption of these procedures was quantified by measuring the power with a wattmeter at the input of the microwave generators and the electricity source of an electrical heater. The standard SD method is associated with a higher CO_2 emission to the atmosphere in terms of environmental impact, with a predicted amount of 238 kg CO_2/kg oil. In comparison, MSD produces less CO_2/kg oil at 95 kg CO_2. These numbers are based on literature sources (Yek et al., 2020), which demonstrate that the combustion of fossil fuels results in the release of around 800 g of CO_2 for every 1 kWh of energy produced from coal or fuel. According to Skendi et al. (2022), MSD also significantly reduces the amount of wastewater generated throughout the extraction procedure from 290 L/kg of essential oil with SD to 85 L/kg.

7.6 INDUSTRIAL APPLICATIONS

In the 1990s, Archimex invented and received a patent for the vacuum microwave hydro-distillation process (Roohinejad et al., 2017). This method combines sequential hoover application with microwave-based selective heating. To rehydrate the dried plant subject matter, a microwave chamber and water (Yousuf et al., 2021) are used. After that, microwave radiation is applied to the plant material to promote the release of natural extracts. By reducing the pressure to a range from 100 to 200 mbar, the volatile azeotropic combination of volatile water and oil can be evacuated from the biological matrix. The entire amount of volatile oil from the plant matter can be extracted sequentially by performing the extraction technique. The system can process up to 100 kg of materials in an hour. The French company Gattefossé (www.gattefosse.com) produces the "original extracts" by microwave distilling the natural water content found in spices, herbs, and fruits without using any water or reagent. These authentic extracts are used in nutraceuticals, cosmetics, and fragrances (Yek et al., 2020).

7.7 CONCLUSION AND PROSPECTS

It has been suggested that microwave-assisted procedures are acceptable replacements for conventional techniques for recovering target components from food items. The electromagnetic field-intensified process has several benefits, including less solvent consumption, a shorter processing time, and a higher extraction yield. This chapter explained how to improve MAE techniques using integrated technology, which combines microwaves with other extraction methods like UAE, EAE, subcritical fluid extraction, SFE, and negative pressure cavitation, as well as using rising more sustainable solvents or methods like ionised fluids, multiphase solvents, deep eutectic solvents, and surfactants that are not in an ionic state at their cloud point temperature. These additions or adjustments to standard MAE techniques offer several additional advantages, such as a further increase in extraction yield, speed, little component deterioration, and energy savings. Despite this, there are still insufficient experimental data on this advancement of MAE approaches. Therefore, it is important that additional research be done to ensure validity and sustainability. Finding new extraction techniques, proposing process improvement, and resolving associated conceptual and technological issues should all receive special consideration. To facilitate the translation of laboratory methods into industrial applications, it is crucial to conduct investigations on the commercial feasibility of implementing these enhanced MAE systems for large-scale recovery of food products. This entails assessing their adaptability, scalability, and economic viability for industrial operations. Until safety is demonstrated, these concepts, particularly those involving solvents, remain up for debate. Nevertheless, it is hoped that the current assessment will spark more research in the field, paving the way for industrial adoption of these cutting-edge methods for the timely commercial processing of food components.

REFERENCES

Addo, P. W., Brousseau, V. D., Morello, V., MacPherson, S., Paris, M., & Lefsrud, M. (2021). Cannabis chemistry, post-harvest processing methods and secondary metabolite profiling: A review. *Industrial Crops and Products*, *170*, 113743.

Alara, O. R., Abdurahman, N. H., & Ukaegbu, C. I. (2021). Extraction of phenolic compounds: A review. *Current Research in Food Science*, *4*, 200–214.

Alexandre, E. M. C., Castro, L. M. G., Moreira, S. A., Pintado, M., & Saraiva, J. A. (2017). Comparison of emerging technologies to extract high-added value compounds from fruit residues: Pressure- and electro-based technologies. *Food Engineering Reviews*, *34*(6), 581–612.

Bagade, S. B., & Patil, M. (2021). Recent advances in microwave-assisted extraction of bioactive compounds from complex herbal samples: A review. *Critical Reviews in Analytical Chemistry*, *51*(2), 138–149.

Baite, T. N., Mandal, B., & Purkait, M. K. (2021). Ultrasound assisted extraction of gallic acid from Ficus auriculata leaves using green solvent. *Food and Bioproducts Processing*, *128*, 1–11.

Bora, H., Kamle, M., Mahato, D. K., Tiwari, P., & Kumar, P. (2020). Citrus essential oils (CEOs) and their applications in food: An overview. *Plants*, *9*(3), 357.

Carneiro, C. N., & Dias, F. D. S. (2021). Multiple response optimisation of ultrasound-assisted procedure for multi-element determination in Brazilian wine samples by microwave-induced plasma optical emission spectrometry. *Microchemical Journal*, *171*, 106857.

Chaves, J. O., De Souza, M. C., Da Silva, L. C., Lachos-Perez, D., Torres-Mayanga, P. C., Machado, A. P. D. F., . . . & Rostagno, M. A. (2020). Extraction of flavonoids from natural sources using modern techniques. *Frontiers in Chemistry*, *8*, 507887.

Chemat, F., Lucchesi, M., & Smadia, J. (2004). U.S. Patent Application No. 10/751,988.

Chemat, F., Smadja, J., & Lucchesie, M. E. (2004). *Solvent free microwave extraction of volatile natural compound*. European Patent EP 1 439218 B.

Chemat, F., Vian, M., & Visioni, F. (2008). *Microwave hydrodiffusion for isolation of natural products*. European Patent EP 1,955,749 A1.

Chemat, F., Vian, M. A., Fabiano-Tixier, A. S., Nutrizio, M., Jambrak, A. R., Munekata, P. E., . . . & Cravotto, G. (2020). A review of sustainable and intensified techniques for extraction of food and natural products. *Green Chemistry*, *22*(8), 2325–2353.

Constantino, D. S., Dias, M. M., Silva, A. M., Faria, J. L., & Silva, C. G. (2022). Intensification strategies for improving the performance of photocatalytic processes: A review. *Journal of Cleaner Production*, *340*, 130800.

Craveiro, A. A., Matos, F. J. A., Alencar, J. W., & Plumel, M. M. (1989). Microwave oven extraction of an essential oil. *Flavour and Fragrance Journal*, *4*(1), 43–44.

Erukainure, O. L., Atolani, O., Muhammad, A., Ravichandran, R., Abarshi, M. M., Katsayal, S. B., . . . & Mesaik, M. A. (2021). Translational suppression of SARS-COV-2 ORF8 protein mRNA as a Viable therapeutic target against COVID-19: Computational studies on potential roles of isolated compounds from Clerodendrum volubile leaves. *Computers in Biology and Medicine*, *139*, 104964.

Fierascu, I., Avramescu, S. M., Sieniawska, E., & Fierascu, R. C. (2023). Ionic liquids with microwave-assisted extraction of natural products. In *Extraction of natural products from agro-industrial wastes* (pp. 35–52). Elsevier.

Ganzler, K., Salgo, A., & Valkó, K. (1986). Microwave extraction: A novel sample preparation method for chromatography. *Journal of Chromatography A*, *371*, 299–306.

Gao, M. Z., Cui, Q., Wang, L. T., Meng, Y., Yu, L., Li, Y. Y., & Fu, Y. J. (2020). A green and integrated strategy for enhanced phenolic compounds extraction from mulberry (Morus alba L.) leaves by deep eutectic solvent. *Microchemical Journal*, *154*, 104598.

Georgiopoulou, I., Tzima, S., Louli, V., & Magoulas, K. (2022). Supercritical CO2 extraction of high-added value compounds from Chlorella vulgaris: Experimental design, modelling and optimisation. *Molecules*, *27*(18), 5884.

Ghazanfari, N., Mortazavi, S. A., Yazdi, F. T., & Mohammadi, M. (2020). Microwave-assisted hydrodistillation extraction of essential oil from coriander seeds and evaluation of their composition, antioxidant and antimicrobial activity. *Heliyon*, *6*(9).

Gholamian, S., Nourani, M., & Bakhshi, N. (2021). Formation and characterisation of calcium alginate hydrogel beads filled with cumin seeds essential oil. *Food Chemistry*, *338*, 128143.

Gong, P., Wang, S., Liu, M., Chen, F., Yang, W., Chang, X., . . . & Chen, X. (2020). Extraction methods, chemical characterisations and biological activities of mushroom polysaccharides: A mini-review. *Carbohydrate Research*, *494*, 108037.

Guo, Y., Baschieri, A., Amorati, R., & Valgimigli, L. (2021). Synergic antioxidant activity of γ-terpinene with phenols and polyphenols enabled by hydroperoxyl radicals. *Food Chemistry*, *345*, 128468.

Hossain, M. S., Sharfaraz, A., Dutta, A., Ahsan, A., Masud, M. A., Ahmed, I. A., . . . & Ming, L. C. (2021). A review of ethnobotany, phytochemistry, antimicrobial pharmacology and toxicology of Nigella sativa L. *Biomedicine & Pharmacotherapy*, *143*, 112182.

Kumar, M., Dahuja, A., Tiwari, S., Punia, S., Tak, Y., Amarowicz, R., Bhoite, A. G., Singh, S., Joshi, S., Panesar, P. S., Saini, R. P., Pihlanto, A., Tomar, M., Sharifi-Rad, J., & Kaur, C. (2021). Recent trends in extraction of plant bioactives using green technologies: A review. *Food Chemistry, 353*, 129431. https://doi.org/10.1016/j.foodchem.2021.129431

Lefebvre, T., Destandau, E., & Lesellier, E. (2021). Selective extraction of bioactive compounds from plants using recent extraction techniques: A review. *Journal of Chromatography A, 1635*, 461770.

Louati, H., Maria, S., Rocci, J. F., & Doumenq, P. (2020). Determination of total hydrocarbons in contaminated soil with "thin layer sorptive extraction coupled with attenuated total reflectance–fourier transform infrared spectroscopy". *Analytical Chemistry, 92*(23), 15344–15351.

Mahanta, B. P., Bora, P. K., Kemprai, P., Borah, G., Lal, M., & Haldar, S. (2021). Thermolabile essential oils, aromas and flavours: Degradation pathways, effect of thermal processing and alteration of sensory quality. *Food Research International, 145*, 110404.

Mandal, V., & Tandey, R. (2016). A critical analysis of publication trends from 2005–2015 in microwave assisted extraction of botanicals: How far we have come and the road ahead. *TrAC Trends in Analytical Chemistry, 82*, 100–108.

More, P. R., Jambrak, A. R., & Arya, S. S. (2022). Green, environment-friendly and sustainable techniques for extraction of food bioactive compounds and waste valorisation. *Trends in Food Science & Technology, 128*, 296–315.

Mukherjee, S., Chouhan, K. B. S., Chandrakar, M., Gupta, P., Lal, K., & Mandal, V. (2022). A cross talk based critical analysis of solvent free microwave extraction to accentuate it as the new normal for extraction of essential oil: An attempt to overhaul the science of distillation through a comprehensive tutelage. *Critical Reviews in Food Science and Nutrition*, 1–23.

Nguyen, E., Jones, O., Kim, Y. H. B., San Martin-Gonzalez, F., & Liceaga, A. M. (2017). Impact of microwave-assisted enzymatic hydrolysis on functional and antioxidant properties of rainbow trout Oncorhynchus mykiss by-products. *Fisheries Science*, 1–15.

Ordoudi, S. A., Papapostolou, M., Nenadis, N., Mantzouridou, F. T., & Tsimidou, M. Z. (2022). Bay Laurel (Laurus nobilis L.) essential oil as a food preservative source: Chemistry, quality control, activity assessment, and applications to olive industry products. *Foods, 11*(5), 752.

Pagano, I., Campone, L., Celano, R., Piccinelli, A. L., & Rastrelli, L. (2021). Green non-conventional techniques for the extraction of polyphenols from agricultural food by-products: A review. *Journal of Chromatography A, 1651*, 462295.

Panzella, L., Moccia, F., Nasti, R., Marzorati, S., Verotta, L., & Napolitano, A. (2020). Bioactive phenolic compounds from agri-food wastes: An update on green and sustainable extraction methodologies. *Frontiers in Nutrition, 7*, 60.

Qiu, S. M., Aweya, J. J., Liu, X., Liu, Y., Tang, S., Zhang, W., & Cheong, K. L. (2022). Bioactive polysaccharides from red seaweed as potent food supplements: A systematic review of their extraction, purification, and biological activities. *Carbohydrate Polymers, 275*, 118696.

Radivojac, A., Bera, O., Zeković, Z., Teslić, N., Mrkonjić, Ž., Bursać Kovačević, D., . . . & Pavlić, B. (2021). Extraction of peppermint essential oils and lipophilic compounds: Assessment of process kinetics and environmental impacts with multiple techniques. *Molecules, 26*(10), 2879.

Rao, M. V., Sengar, A. S., Sunil, C. K., & Rawson, A. (2021). Ultrasonication-A green technology extraction technique for spices: A review. *Trends in Food Science & Technology, 116*, 975–991.

Rasekh, M., Karami, H., Wilson, A. D., & Gancarz, M. (2021). Classification and identification of essential oils from herbs and fruits based on a MOS electronic-nose technology. *Chemosensors*, *9*(6), 142.

Roohinejad, S., Koubaa, M., Barba, F. J., Leong, S. Y., Khelfa, A., Greiner, R., & Chemat, F. (2017). Extraction methods of essential oils from herbs and spices. *Essential Oils in Food Processing: Chemistry, Safety and Applications*, *1*, 21–55.

Saha, N., Sarkar, B., & Sen, K. (2022). Aqueous biphasic systems: A robust platform for green extraction of biomolecules. *Journal of Molecular Liquids*, 119882.

Sahraoui, N., AbertVian, M., Bornard, I., Boutekedjiret, C., & Chemat, F. (2008). Improved microwave steam distillation apparatus for isolation of essential oils, comparison with conventional steam distillation. *Journal of Chromatography A*, *1210*(2008), 229–233.

Sarker, T. R., Pattnaik, F., Nanda, S., Dalai, A. K., Meda, V., & Naik, S. (2021). Hydrothermal pretreatment technologies for lignocellulosic biomass: A review of steam explosion and subcritical water hydrolysis. *Chemosphere*, *284*, 131372.

Sharma, S., Barkauskaite, S., Jaiswal, A. K., & Jaiswal, S. (2021). Essential oils as additives in active food packaging. *Food Chemistry*, *343*, 128403.

Shen, F., Song, Z., Xie, P., Li, L., Wang, B., Peng, D., & Zhu, G. (2021). Polygonatum sibiricum polysaccharide prevents depression-like behaviors by reducing oxidative stress, inflammation, and cellular and synaptic damage. *Journal of Ethnopharmacology*, *275*, 114164.

Skendi, A., Irakli, M., Chatzopoulou, P., Bouloumpasi, E., & Biliaderis, C. G. (2022). Phenolic extracts from solid wastes of the aromatic plant essential oil industry: Potential uses in food applications. *Food Chemistry Advances*, 100065.

Stashenko, E. E., Jaramillo, B. E., & Martinez, J. R. (2004a). Comparison of different extraction methods for the analysis of volatile secondary metabolites of Lippia alba (Mill.) N.E. Brown, grown in Colombia, and evaluation of its in vitro antioxidant activity. *Journal of Chromatography A*, *1025*, 93–103.

Stashenko, E. E., Jaramillo, B. E., & Martınez, J. R. (2004b). Analysis of volatile secondary metabolites from Colombian *Xylopiaaromatica* (Lamarck) by different extraction and headspace methods and gas chromatography". *Journal of Chromatography A*, *1025*(1), 105–113.

Stephane, F. F. Y., & Jules, B. K. J. (2020). Terpenoids as important bioactive constituents of essential oils. In *Essential Oils-Bioactive Compounds, New Perspectives and Applications* (pp. 1–15). IntechOpen.

Tasa, H., Imani, A., Moghanlou, K. S., Nazdar, N., & Moradi-Ozarlou, M. (2020). Aflatoxicosis in fingerling common carp (Cyprinus carpio) and protective effect of rosemary and thyme powder: Growth performance and digestive status. *Aquaculture*, *527*, 735437.

Yek, P. N. Y., Peng, W., Wong, C. C., Liew, R. K., Ho, Y. L., Mahari, W. A. W., . . . & Lam, S. S. (2020). Engineered biochar via microwave CO2 and steam pyrolysis to treat carcinogenic Congo red dye. *Journal of Hazardous Materials*, *395*, 122636.

Yousuf, B., Wu, S., & Siddiqui, M. W. (2021). Incorporating essential oils or compounds derived thereof into edible coatings: Effect on quality and shelf life of fresh/fresh-cut produce. *Trends in Food Science & Technology*, *108*, 245–257.

Yu, C., Zhang, J., & Wang, T. (2021). Star anise essential oil: Chemical compounds, antifungal and antioxidant activities: A review. *Journal of Essential Oil Research*, *33*(1), 1–22.

Zhang, Z., Pan, S. Y., Li, H., Cai, J., Olabi, A. G., Anthony, E. J., & Manovic, V. (2020). Recent advances in carbon dioxide utilisation. *Renewable and Sustainable Energy Reviews*, *125*, 109799.

Zhao, S., Zhang, Y., Lin, J., Wang, A., Wang, Y., Zhang, Y., . . . & Song, R. (2023). Chemical index components and quality control of traditional chinese medicine: "Never change a winning team"?-A case study of volatile oil from Bupleuri radix. *Journal of Pharmaceutical and Biomedical Analysis*, 115618.

Zhu, S. C., Shi, M. Z., Yu, Y. L., & Cao, J. (2022). Simultaneous extraction and enrichment of alkaloids from lotus leaf by in-situ cloud point-reinforced ionic liquid assisted mechanochemical extraction technology. *Industrial Crops and Products*, *183*, 114968.

Zhu, Z., He, J., Liu, G., Barba, F. J., Koubaa, M., Ding, L., Bals, O., Grimi, N., & Vorobiev, E. (2016). Recent insights for the green recovery of inulin from plant food materials using non-conventional extraction technologies: A review. *Innovative Food Science & Emerging Technologies*, *33*, 1–9.

Zia, S., Khan, M. R., Shabbir, M. A., Aslam Maan, A., Khan, M. K. I., Nadeem, M., . . . & Aadil, R. M. (2022). An inclusive overview of advanced thermal and nonthermal extraction techniques for bioactive compounds in food and food-related matrices. *Food Reviews International*, *38*(6), 1166–1196.

8 Nutraceutical Constituents and Properties and Applications of Essential Oil

Anamika Chauhan

8.1 INTRODUCTION

An overview of essential oils will be provided in this introduction, with a focus on the growing interest in using them as nutraceuticals. Essential oils are obtained from plants using a variety of extraction techniques and have been used for their therapeutic and medical effects since ancient times. Due to its numerous health advantages, there has been an increasing interest in examining the potential of using essential oils as nutraceuticals in recent years. The fragrant molecules that make up essential oils are highly compressed and come from a range of plant parts, including leaves, flowers, stems, and roots. Steam distillation, expression, and solvent extraction are a few techniques for removing them (Bakkali et al., 2008). These oils contain a variety of volatile chemical compounds that contribute to their distinctive aroma and advantageous effects, including terpenes, phenols, aldehydes, and ketones (Kamatou et al., 2013). A "nutraceutical" is a natural product that, when consumed as a regular element of a diet, offers physiological benefits and fosters health (Shahidi & Ambigaipalan, 2015). Nutraceuticals encompass a wide range of bioactive compounds found in food sources, including essential oils. The potential health advantages and therapeutic qualities of essential oils have stoked interest in them as nutraceuticals. The antibacterial, antioxidant, anti-inflammatory, and analgesic properties of these oils have also been noted (Burt, 2004 and Edris, 2007). Essential oils, for instance, are appealing for applications in food preservation because they can limit the growth of food-borne pathogens and increase the shelf life of perishable goods (Friedman et al., 2002). To support numerous areas of health and well-being, essential oils have also been employed in dietary supplements (Setzer, 2009). The versatility of essential oils extends to aromatherapy, where they are used for their calming,

DOI: 10.1201/9781032634180-8

uplifting, and stress-reducing effects (Koulivand et al., 2013). Moreover, essential oils have shown promising potential in addressing specific health concerns, such as digestive issues, headaches, and stress-related disorders (Perry et al., 2000, Shrivastava et al., 2010 and Sasannejad et al., 2012). Hence, essential oils have garnered attention as potential nutraceuticals due to their diverse properties and applications. Their rich chemical composition and bioactive constituents offer promising prospects for improving food safety, enhancing health, and supporting overall well-being. This chapter aims to explore the nutraceutical constituents, properties, applications, and benefits of essential oils, shedding light on their potential as valuable natural products in the field of nutrition and health.

8.2 DEFINITION AND SIGNIFICANCE OF ESSENTIAL OILS

Essential oils are highly concentrated volatile liquids that come from various plant parts, such as leaves, flowers, stems, and roots. Methods like steam distillation, expression, or solvent extraction are used to extract the aromatic compounds that give the plants their characteristic flavour and therapeutic properties (Bakkali et al., 2008 and Kamatou et al., 2013). The importance of essential oils is seen in the wide range of biological activity they exhibit and the possible applications they have. Due to their therapeutic, aromatic, and cosmetic benefits, these oils have been utilized for centuries. They have a complicated chemical make-up that includes terpenes, phenols, aldehydes, ketones, and other kinds of chemicals (Edris, 2007). Since essential oils' volatile components contribute to their aromatic properties, they are highly prized for use in aromatherapy, perfumery, and other natural scent applications (Bakkali et al., 2008) Additionally, scientific study into the possible health advantages of essential oils has been sparked by the medicinal qualities of these substances.

The chemical components included in essential oils have been related to a wide range of biological functions. They might, for instance, possess antibacterial properties that enable them to prevent the growth of many diseases caused by bacteria, fungi, and viruses (Burt, S., 2004 and Friedman et al., 2002). Additionally, essential oils have shown antioxidant activity, which raises the possibility of them scavenging free radicals and guarding against oxidative stress (Kamatou et al., 2013).

The potential application of essential oils in the food, agricultural, and medical sectors increase their worth. Essential oils are used as natural flavourings and preservatives in the food industry because of their antibacterial qualities and capacity to improve the sensory qualities of food products (Burt, S., 2004). They can also be employed in the formulation of natural pesticides and insecticides in agriculture. In medicine, essential oils are increasingly being explored for their potential therapeutic uses. They have shown promising results in areas such as pain management, skin care, respiratory health, and stress reduction (Edris, A. E., 2007 and Koulivand et al., 2013). Essential oils are utilized in complementary and alternative medicine practices, including aromatherapy, where they are used for their calming, soothing, and mood-enhancing effects.

8.3 EMERGING INTEREST IN ESSENTIAL OILS AS NUTRACEUTICALS

Essential oils, derived from various plants, have gained significant attention in recent years due to their potential health benefits and functional properties as displayed in Tables 8.1–8.6. The use of essential oils as nutraceuticals, which are natural compounds found in food sources that offer physiological benefits and promote overall health when consumed as part of a regular diet, has been an area of growing interest. The following are some key points highlighting the emerging interest in essential oils as nutraceuticals:

Bioactive substances: Terpenes, phenolics, and flavonoids are just a few examples of the bioactive substances abundant in essential oils that may have health advantages. They have "antioxidant, antiinflammatory, antimicrobial, and anticancer properties, among others", according to the literature (Azimi et al., 2020).

Antimicrobial activity: Essential oils have proven to be effective against a variety of diseases, including "bacteria, fungi, and viruses". They have been researched for their possible application in food preservation and as all-natural antibacterial agents that can replace synthetic ones (Friedman et al., 2002)

Antioxidant potential: A lot of essential oils have significant antioxidant activity, which helps guard against oxidative stress and damage brought on by free radicals. The antioxidants in essential oils support their potential for enhancing cardiovascular health, lowering the risk of chronic diseases, and slowing the ageing process.

Anti-inflammatory effects: Essential oils have anti-inflammatory characteristics that may reduce the signs and symptoms of illnesses linked to inflammation. They have been investigated in regard to how they might

TABLE 8.1
Nutraceutical constituents of essential oils.

Nutraceutical constituent	Potential health benefits	References
Terpenes (e.g. limonene, pinene, linalool)	Antioxidant, anti-inflammatory, antimicrobial, anticancer effects	Azimi et al. (2020)
Phenols (e.g. thymol, eugenol, carvacrol)	Antioxidant, anti-inflammatory, antimicrobial effects	Azimi et al. (2020)
Aldehydes (e.g. citral, citronellal, cinnamaldehyde)	Antimicrobial, antifungal, calming effects	Bakkali et al. (2008)
Ketones (e.g. camphor, menthone, carvone)	Stimulating, uplifting, antispasmodic effects	Bakkali et al. (2008)
Esters	Calming, sedative effects	Bakkali et al. (2008)

TABLE 8.2
Chemical composition of essential oils.

Chemical component	Source	References
Limonene	Citrus fruits	Bakkali et al. (2008)
Linalool	Lavender, chamomile	Azimi et al. (2020)
Menthol	Peppermint, spearmint	Bakkali et al. (2008)
Eugenol	Clove, cinnamon	Azimi et al. (2020)
Geraniol	Rose, geranium	Bakkali et al. (2008)
α-Pinene	Pine, rosemary	Bakkali et al. (2008)
β-Caryophyllene	Black pepper, cloves	Azimi et al. (2020)
Citral	Lemongrass, lemon	Bakkali et al. (2008)
Camphor	Camphor tree, rosemary	Azimi et al. (2020)

TABLE 8.3
Common chemical constituents and their properties.

State	Properties	Potential health benefits
Limonene	Aromatic, volatile, hydrophobic	Antioxidant, anti-inflammatory, anticancer properties
Pinene	Aromatic, volatile, hydrophobic	Anti-inflammatory, bronchodilator, antimicrobial effects
Linalool	Aromatic, volatile, hydrophobic	Sedative, anxiolytic, antimicrobial, anti-inflammatory effects
Menthol	Aromatic, volatile, hydrophobic	Analgesic, cooling, respiratory benefits
Geraniol	Aromatic, volatile, hydrophobic	Antioxidant, antimicrobial, anti-inflammatory effects
Caryophyllene	Aromatic, volatile, hydrophobic	Anti-inflammatory, analgesic, antioxidant properties
Citral	Aromatic, volatile, hydrophobic	Antimicrobial, antifungal, antioxidant effects

Source: Bakkali et al. (2008) and Azimi et al. (2020).

be used to treat inflammatory conditions such as inflammatory bowel disease and arthritis (Bakkali et al., 2008)

Digestive health: Studies have shown that several essential oils can improve digestive health. They might help with the treatment of illnesses, like irritable bowel syndrome, and can improve digestion, ease gastrointestinal discomfort, and aid in digesting.

Mental health: Essential oils are known for their aromatherapeutic qualities, which have been associated with improved mood, reduced stress,

TABLE 8.4
Phenols, properties, and health benefits.

Phenol	Properties	Potential health benefits	References
Thymol	Aromatic, antimicrobial, antioxidant	Antimicrobial, antifungal, antioxidant properties	Bakkali et al. (2008) and Azimi et al. (2020)
Eugenol	Aromatic, antimicrobial, antioxidant	Analgesic, antimicrobial, antioxidant properties	Bakkali et al. (2008) and Azimi et al. (2020)
Carvacrol	Aromatic, antimicrobial, antioxidant	Antimicrobial, antifungal, antioxidant properties	Bakkali et al. (2008) and Azimi et al. (2020)
Cinnamaldehyde	Aromatic, antimicrobial, antioxidant	Antimicrobial, antifungal, antioxidant properties	Bakkali et al. (2008) and Azimi et al. (2020)
Rosmarinic acid	Antioxidant, anti-inflammatory	Antioxidant, anti-inflammatory, neuroprotective effects	Serafini et al. (2010)
Curcumin	Antioxidant, anti-inflammatory	Antioxidant, anti-inflammatory, anticancer properties	Chainani-Wu (2003)

TABLE 8.5
Aldehydes, properties, and potential health benefits.

Aldehyde	Properties	Potential health benefits
Citral	Aromatic, antimicrobial, calming	Antimicrobial, antifungal, calming effects
Citronellal	Aromatic, antimicrobial, soothing	Antimicrobial, insect-repellent, soothing properties
Cinnamaldehyde	Aromatic, antimicrobial, warming	Antimicrobial, antifungal, warming effects
Benzaldehyde	Aromatic, antimicrobial, soothing	Antimicrobial, soothing properties
Vanillin	Aromatic, antioxidant, calming	Antioxidant, antimicrobial, calming effects
Furfural	Aromatic, antimicrobial, antioxidant	Antimicrobial, antioxidant properties

Source: Bakkali et al. (2008) and Azimi et al. (2020).

and relaxation. Because of their propensity to reduce anxiety and improve mood, they are frequently utilized in aromatherapy (Serafini et al., 2010).

Nutritional support: Essential oils can help with nutrition by supplying vitamins, minerals, and other healthy substances. Citrus essential oils,

TABLE 8.6
Ketones, properties, and potential health benefits.

Ketone	Properties	Potential health benefits
Camphor	Aromatic, antimicrobial, decongestant	Antimicrobial, decongestant, analgesic effects
Menthone	Aromatic, cooling	Cooling, analgesic, insect-repellent properties
Carvone	Aromatic, digestive aid	Digestive aid, antimicrobial, anti-inflammatory effects
Pulegone	Aromatic, insect-repellent	Insect-repellent, antimicrobial, analgesic effects
Menthone	Aromatic, cooling	Cooling, analgesic, antifungal properties
2-Heptanone	Aromatic, antimicrobial	Antimicrobial, sedative properties

Source: Bakkali et al. (2008) and Azimi et al. (2020).

for instance, are rich in vitamin C, whereas oils made from spices and herbs may include phytonutrients that have potential health advantages.

Delivery systems: New delivery technologies, such microencapsulation and nano emulsions, have made it easier to incorporate essential oils into different food and beverage items. These developments are intended to increase the components of essential oils' stability, bioavailability, and controlled release, making them better suited for use in nutraceutical products (Perry & Perry, 2006). Despite the positive signs of the growing interest in essential oils as nutraceuticals, more study is required to completely comprehend their mechanisms of action, ideal dosages, and potential drug interactions. Additionally, quality control and standardization of essential oils are crucial to ensure their safety and efficacy. Hence, the increasing exploration of essential oils as nutraceuticals presents exciting possibilities for their use in promoting health and well-being. However, it is important to consult with healthcare professionals or experts in the field before incorporating essential oils into your diet or wellness routine.

8.4 PROPERTIES OF ESSENTIAL OILS

Aromatic properties: Essential oils are highly aromatic substances that contain volatile compounds responsible for their distinct fragrance. These compounds can stimulate the olfactory system and provide a pleasant aroma (Lis-Balchin, 2006), as shown in Tables 8.7–8.10.

Antimicrobial properties: Many essential oils possess antimicrobial properties and can inhibit the growth of bacteria, fungi, and viruses. These properties make essential oils useful for sanitization, disinfection, and preservation purposes (Hammer et al., 1999).

Anti-inflammatory properties: Some essential oils exhibit anti-inflammatory effects and can help reduce inflammation in the body. These oils may

TABLE 8.7
Antimicrobial properties.

Essential oils	Microorganisms targeted	Potential applications
Tea Tree	Bacteria, fungi, viruses	Acne treatment, wound healing, oral hygiene
Eucalyptus	Bacteria, fungi, viruses	Respiratory infections, skin infections, oral health
Peppermint	Bacteria, fungi, viruses	Oral health, respiratory infections, skin infections
Cinnamon	Bacteria, fungi, viruses	Food preservation, oral health, fungal infections
Oregano	Bacteria, fungi, viruses	Food preservation, respiratory infections
Lemongrass	Bacteria, fungi, viruses	Skin infections, oral health, insect repellent

Source: Carson et al. (2006) and Hammer et al. (1999).

TABLE 8.8
Antioxidant properties.

Essential oils	Antioxidant compounds	Potential applications	References
Rosemary	Rosmarinic acid, carnosic acid	Food preservation, skin health, neuroprotection	González-Trujano et al. (2017)
Lavender	Linalool, linalyl acetate	Skin health, stress relief, neuroprotection	Viljoen et al. (2012) and Lis-Balchin (2006)
Frankincense	Boswellic acids	Anti-inflammatory, skin health, anti-ageing	Siddiqui (2011) and Al-Yasiry et al. (2016)
Thyme	Thymol, carvacrol	Food preservation, antimicrobial, skin health	Teixeira et al. (2013) and Soković et al. (2009)
Cinnamon	Cinnamaldehyde, eugenol	Food preservation, anti-diabetic, anti-inflammatory	Rao et al. (2014) and Shan et al. (2018)
Lemon	Limonene, citral	Immune support, detoxification, skin health	Sadiki et al. (2017) and Kim et al. (2020)

provide relief from inflammatory conditions such as arthritis, skin irritations, and respiratory issues (Peana et al., 2002).

Antioxidant properties: Many essential oils contain compounds with antioxidant properties, which help protect cells from oxidative stress and damage caused by free radicals. These antioxidants can support overall health and may have anti-ageing effects (Hussain et al., 2008).

Relaxing and calming properties: Certain essential oils have soothing and sedative effects on the body and mind. They can support stress reduction, relaxation, and better sleep (Setzer et al., 2009).

TABLE 8.9
Anti-inflammatory properties.

Essential oils	Anti-inflammatory compounds	Potential applications	References
Chamomile	α-Bisabolol, matricin	Skin inflammation, digestive disorders	Shrivastava et al. (2010) and Aguirre et al. (2020)
Lavender	Linalool, linalyl acetate	Skin inflammation, stress relief, pain management	Liao et al. (2019) and Barocelli et al. (2012)
Frankincense	Boswellic acids	Arthritis, skin inflammation, respiratory disorders	Siddiqui (2011) and Al-Yasiry et al. (2016)
Eucalyptus	Eucalyptol (1,8-cineole)	Respiratory inflammation, muscle/joint pain	Juergens et al. (2003) and Silva et al. (2003)
Rosemary	Rosmarinic acid, carnosic acid	Neuroinflammation, digestive disorders	Choi et al. (2013) and Anwar et al. (2019)
Turmeric	Curcumin	Joint inflammation, digestive disorders	Aggarwal et al. (2007) and Hewlings et al. (2017)

TABLE 8.10
Analgesic properties.

Essential oils	Analgesic compounds	Potential applications	References
Peppermint	Menthol	Headaches, muscle pain, dental pain, postoperative pain	Göbel et al. (2016) and Khanna et al. (2012)
Lavender	Linalool, linalyl acetate	Headaches, migraines, menstrual cramps, general pain	Koulivand et al. (2013) and Seyedian et al. (2020)
Eucalyptus	Eucalyptol (1,8-cineole)	Muscle/joint pain, headaches, postoperative pain	Juergens et al. (2003) and Silva et al. (2003)
Rosemary	Rosmarinic acid, carnosic acid	Headaches, muscle pain, arthritis	Satou et al. (2019) and Prusinowska et al. (2019)
Clove	Eugenol	Dental pain, arthritis, muscle pain	Da Silva et al. (2017) and Bhalla et al. (2013)
Frankincense	Boswellic acids	Arthritis, general pain	Siddiqui (2011) and Al-Yasiry et al. (2016)

8.5 APPLICATIONS OF ESSENTIAL OILS AS NUTRACEUTICALS

8.5.1 Food Preservation and Flavouring

Food preservation and flavouring are important applications of essential oils. Essential oils possess antimicrobial properties that can help inhibit the growth

TABLE 8.11
Dietary supplements.

Dietary supplements	Role	Food sources	References
Vitamin D	Plays a crucial role in calcium absorption and bone health, immune function, and modulation of cell growth and differentiation	Fatty fish (such as salmon and mackerel), fortified dairy products, sunlight exposure	Holick (2007)
Omega-3 fatty acids	Essential for brain health, reducing inflammation, and promoting heart health	Fatty fish (such as salmon, tuna, and sardines), walnuts, flaxseeds, chia seeds	Mozaffarian & Rimm (2006)
Probiotics	Promote a healthy gut microbiota, improve digestion, support immune function, and may have potential benefits for various health conditions	Fermented foods (yogurt, kefir, sauerkraut, kimchi)	Hill et al. (2014)
Calcium	Essential for strong bones and teeth, proper muscle function, nerve transmission, and blood clotting	Dairy products, leafy green vegetables, fortified foods	Weaver & Plawecki (1994)

of spoilage and pathogenic microorganisms in food, thereby extending its shelf life. Additionally, essential oils can contribute to the flavour and aroma profile of foods, enhancing their sensory appeal (see Tables 8.11 and 8.12).

Food Preservation:

Antimicrobial properties: Essential oils, such as oregano, thyme, cinnamon, and clove, have been found to exhibit broad-spectrum antimicrobial activity against various food-borne pathogens and spoilage microorganisms (Burt, S., 2004).

Preservation techniques: Essential oils can be used in different preservation techniques, including direct addition to foods, incorporation into edible films or coatings, and packaging systems (Singh et al., 2019).

Mechanisms of action: Essential oils have antibacterial properties in part because they can damage microbial cell membranes, limit enzyme function, alter cellular metabolism, and act as antioxidants (Nazzaro et al., 2013). Applications include the successful preservation of meat, poultry, seafood, fruits, vegetables, dairy products, and baked goods using essential oils (Burt & Reinders, 2003 and Prakash et al., 2015)

TABLE 8.12
Other therapeutic applications.

Therapeutic applications	Details	Authors
Stress and anxiety relief	By encouraging calm and relaxation, aromatherapy can help lower stress and anxiety levels. It has been demonstrated that several essential oils, such as "lavender and bergamot", have calming benefits.	Perry & Perry (2006)
Sleep improvement	Because they contain calming qualities, several essential oils, such "lavender and chamomile", can help with insomnia symptoms and improve the quality of sleep.	Setzer (2009)
Pain management	Numerous types of discomfort, such as headaches, muscle aches, and menstrual cramps, can be treated using aromatherapy. Essential oils like peppermint and eucalyptus have analgesic and anti-inflammatory properties.	Buckle (2015)
Mood enhancement	Aromatherapy can positively impact mood and emotional well-being. Citrus essential oils, such as orange and lemon, are known for their uplifting and energizing effects, while floral oils like rose and ylang-ylang have mood-stabilizing properties.	Tisserand & Young (2014)
Respiratory health	Essential oils like "eucalyptus and tea tree" have expectorant and decongestant properties, making them useful in managing respiratory conditions such as colds, coughs, and sinus congestion.	Lis-Balchin (2006)

Flavouring:

Enhancing flavour and aroma: Essential oils are full of volatile chemicals that give food its distinctive flavours and fragrances. For instance, orange, lemon, and lime essential oils are frequently used to impart acidic and reviving notes to sauces, sweets, and beverages (Delaquis et al., 2002). Essential oils can be used as natural substitutes for artificial flavours, giving food products rich and realistic flavour profiles (Baser et al., 2008).

Dosage and formulation: Because essential oils are so potent, it is best to use them sparingly to prevent overpowering flavours. The stability and regulated release of essential oils in food systems can be maintained with the use of suitable formulation techniques, such as microencapsulation (Roopesh et al., 2015). Compliance with regulatory guidelines and specifications is crucial for food safety (Ribeiro-Santos et al., 2018).

8.5.2 Inhibition of Food-borne Pathogens

Salmonella enterica:

Essential oils: Essential oils of oregano, thyme, cinnamon, and clove have shown strong inhibitory effects against *Salmonella enterica* (Skandamis et al., 2001).

Mechanism of action: Essential oils disrupt the cell membranes of *Salmonella*, leading to cell death and inhibition of growth (Gutierrez et al., 2009).

Escherichia coli O157:H7:

Essential oils: Essential oils of oregano, thyme, cinnamon, and clove have demonstrated effective inhibition of *Escherichia coli O157:H7* (Skandamis et al., 2001).

Mechanism of action: Essential oils can disrupt the cell membrane integrity of *E. coli*, leading to cell leakage and inhibition of growth (Gutierrez et al., 2009).

Listeria monocytogenes:

Essential oils: Essential oils of thyme, oregano, rosemary, and cinnamon have been shown to exhibit strong inhibitory activity against *Listeria monocytogenes* (Djenane et al., 2012 and Chorianopoulos et al., 2004).

Mechanism of action: Essential oils can disrupt the cell membrane of *Listeria monocytogenes*, leading to cell death and inhibition of growth (Djenane et al., 2012).

8.6 AROMATHERAPY

- *Definition*: "Aromatherapy is a holistic healing practice that utilizes essential oils extracted from plants to promote physical, mental, and emotional well-being" (Tisserand & Young, 2014).
- *Essential oils*: Aromatherapy relies on the use of essential oils, which are highly concentrated plant extracts that capture the aromatic compounds and therapeutic properties of plants (Lawless, 2013).
- *Inhalation*: Inhalation is the most popular way to apply essential oils in aromatherapy. Direct inhalation, diffuser use, or inhaling the steam created by mixing essential oils to hot water are all methods of inhalation (Buckle, 2015).
- *Effect on mind and emotions*: Aromatherapy has a big impact on the mind and emotions. Numerous essential oils have particular properties that can aid in boosting mood, relaxation, and mental clarity (Perry & Perry, 2006).
- The bodily health of a person may benefit from aromatherapy. Some essential oils have antimicrobial properties that support the immune system's function and aid to lessen pain and inflammation (Lis-Balchin, 2006).

- *Application methods*: Essential oils can be applied topically via massage, baths, compresses, and skincare items in addition to being inhaled. Essential oils should be used with caution and in accordance with safety measures to prevent skin irritation (Price & Price, 2012).
- *Research and support*: Although aromatherapy has been used for many years, scientific research on its effectiveness is still in its infancy. Studies have revealed encouraging advantages in areas like stress reduction, improved sleep, pain management, and mood enhancement (Shahidi & Ambigaipalan, 2015).

8.7 BENEFITS OF ESSENTIAL OILS AS NUTRACEUTICALS

8.7.1 Health Benefits

- *Stress reduction*: Aromatherapy has been shown to help reduce stress levels and promote relaxation. Essential oils such as "lavender, chamomile, and ylang-ylang" have calming properties that can help alleviate stress and anxiety (Perry & Perry, 2006).
- *Improved sleep*: Certain essential oils, including lavender and bergamot, have sedative effects and can promote better sleep. Aromatherapy using these oils has been shown to improve sleep quality and reduce insomnia symptoms (Setzer, 2009).
- *Pain relief*: Essential oils like peppermint, eucalyptus, and lavender have analgesic properties and can help alleviate pain. Aromatherapy massage or topical application of these oils may provide relief from headaches, muscle aches, and joint pain (Buckle, 2015).
- *Enhanced mood*: The effects of aromatherapy on mood and mental health can be beneficial. Essential oils such as "citrus oils (orange, lemon) and floral oils (rose, jasmine) have uplifting and mood-enhancing effects" (Tisserand & Young, 2014).
- *Improved respiratory health*: Inhalation of essential oils like eucalyptus, tea tree, and peppermint can help clear congestion, soothe coughs, and promote easier breathing. These oils have expectorant and decongestant properties that benefit respiratory health (Lis-Balchin, 2006).
- *Enhanced cognitive function*: Some essential oils, such as "rosemary and peppermint", have been associated with improved cognitive function, memory, and mental alertness. Aromatherapy using these oils may help enhance focus and concentration (Moss et al., 2008).
- *Immune system support*: Certain essential oils, including "tea tree, eucalyptus, and lemon", possess antimicrobial properties that can support the immune system and help fight against bacteria, viruses, and fungi (Hammer et al., 1999).
- *Skin health*: Aromatherapy using essential oils like "tea tree, lavender, and chamomile" can benefit skin health. These oils have anti-inflammatory, antiseptic, and soothing properties that can help with various skin

TABLE 8.13
Benefits of essential oils as nutraceuticals.

Benefits of essential oils as nutraceuticals	Details	References
Antioxidant activity	Essential oils, such as oregano, rosemary, and thyme, exhibit potent antioxidant properties. They can aid in scavenging free radicals and guarding against oxidative cell damage.	Bakkali et al. (2008)
Anti-inflammatory effects	Some essential oils, including frankincense, chamomile, and lavender, possess anti-inflammatory properties. They can help reduce inflammation and alleviate symptoms of inflammatory conditions.	De Cássia da Silveira e Sá et al. (2013)
Antimicrobial activity	Many essential oils, such as "tea tree, eucalyptus, and cinnamon", exhibit broad-spectrum antimicrobial activity. They can help inhibit the growth of bacteria, fungi, and viruses	Hammer et al. (1999)
Digestive support	Essential oils like "peppermint, ginger, and fennel" have been used traditionally to support digestion. They can help alleviate digestive discomfort, promote healthy digestion, and relieve symptoms such as bloating and indigestion.	Grigoleit & Grigoleit (2005)
Mood enhancement	Aromatherapy with essential oils, such as "lavender, bergamot, and citrus oils", can have mood-enhancing effects. They can help reduce stress, anxiety, and promote relaxation and positive emotions.	Perry & Perry (2006)
Skin health		Carson et al. (2006)

conditions, including acne, eczema, and minor irritations (Carson et al., 2006) (see Table 8.13).

8.7.2 Potential Therapeutic Uses

Relaxation and stress relief: Essential oils like lavender, chamomile, and bergamot have soothing properties that can help promote "relaxation, reduce stress, and improve sleep quality" (Koulivand et al., 2013 and Wang & Kulkarni, 2020).

Respiratory health: Essential oils such as "eucalyptus, peppermint, and tea tree" have expectorant and decongestant properties that can help relieve respiratory congestion, coughs, and sinusitis (Juergens et al., 1998 and Serafino et al., 2008).

Pain management: Certain essential oils, including "peppermint, ginger, and lavender", possess analgesic properties that can help alleviate pain and discomfort associated with headaches, muscle aches, and arthritis (Göbel et al., 1994 and Koulivand et al., 2013).

Skin care: Essential oils like "tea tree, lavender, and chamomile" have antimicrobial and anti-inflammatory properties that can be beneficial for skin conditions such as acne, eczema, and dermatitis (Carson et al., 2006).

Digestive support: Essential oils such as peppermint, ginger, and fennel can help support digestion, alleviate digestive discomfort, and promote healthy gut function (Badgujar et al., 2014).

Mood enhancement: Citrus essential oils like lemon, orange, and grapefruit have uplifting and mood-enhancing effects that can help improve mood, reduce anxiety, and boost energy levels (Kuroda et al., 2005 and Sayorwan et al., 2013).

Cognitive function: Rosemary and peppermint essential oils have been associated with improved cognitive function, memory, and focus. Inhalation or topical application of these oils may enhance mental alertness and concentration (Moss et al., 2003, 2009).

Immune support: Some essential oils, including "tea tree, eucalyptus, and thyme", have antimicrobial properties that can support the immune system and help fight against bacteria, viruses, and fungi (Soković et al., 2010 and Budzyńska et al., 2011).

8.8 CONCLUSION

- As nutraceuticals, essential oils have drawn a lot of interest and attention because of their unique chemical make-up and potential health advantages.
- Terpenes, phenols, aldehydes, and ketones are some of the nutraceutical components that provide essential oils their therapeutic benefits and biological effects.
- Essential oils have a number of beneficial characteristics, such as analgesic, antibacterial, antioxidant, anti-inflammatory, and digestive support, which can improve general well-being and address particular health issues.
- The capacity of these oils to taste food, preserve it, and prevent the growth of germs that cause food illness makes them valuable in the food industry. Essential oils can improve nutrition and have additional good impacts on health when they are included in a balanced diet.
- Aromatherapy, or the therapeutic application of essential oils, has long been recognized for its relaxing, stress-relieving, and mood-lifting effects.
- Despite the potential benefits of essential oils, it is crucial to utilize them carefully, taking into account each user's level of sensitivity, the appropriate dilution, and expert guidance.

Also contributing to the increased interest in essential oils as nutraceuticals are their diverse chemical compositions and potential health benefits. These oils offer a range of properties that support their therapeutic usage for things like pain treatment, gastrointestinal support, skin care, and mental acuity improvement. They also serve as dietary supplements, aromatherapy, and food preservatives, all of which have medical uses. However, when utilizing essential oils for particular health issues, it is crucial to use caution and seek out expert counsel.

8.8.1 Summary of the Importance of Essential Oils as Nutraceuticals

Because of the variety of chemicals they contain and the possible health advantages, essential oils are significant as nutraceuticals. Numerous bioactive substances found in essential oils have the potential to improve general health and to target particular health issues. They are useful for a range of therapeutic applications because they have analgesic, antibacterial, antioxidant, anti-inflammatory, and digestive support characteristics.

Essential oils have drawn attention due to their potential for flavouring, preserving food, and thwarting the spread of food-borne diseases, providing organic substitutes for artificial additives. To add more nutritional support and advance some health advantages, they can be utilized as dietary supplements. Aromatherapy, which employs essential oils for therapeutic purposes, has demonstrated potential to aid in stress relief, mood enhancement, and support for mental health. It is necessary to use essential oils correctly and securely, though. To guarantee their efficient and secure usage, proper dilution, awareness of individual sensitivities, and consulting experienced experts are crucial.

In conclusion, because of the wide chemical make-up and possible health benefits of essential oils, they have become important nutraceuticals. They play a significant role among natural therapies for enhancing general health and well-being because of their characteristics, uses in food preservation and flavouring, use as dietary supplements, and therapeutic applications like aromatherapy.

8.8.2 Need for Further Research and Optimization in Nutraceutical Formulations

Even though essential oils have a bright future as nutraceuticals, more work needs to be done to improve their formulation and use. Several aspects that need addressing are:

- *Standardization of composition*: The chemical composition of essential oils might vary depending on plant species, region of origin, and extraction techniques. To guarantee constant quality and effectiveness in nutraceutical formulations, essential oil composition standards must be established.

- Studies on safety and toxicity are still needed to determine the essential oils' long-term safety and toxicity profiles when taken as nutraceuticals. Finding safe dosage ranges, any drug interactions, and any negative effects related to long-term or high-dose use are all included in this.
- *Bioavailability and formulation optimization*: To maximize the therapeutic effects of essential oils, it is vital to increase the bioavailability of their active components. To increase the body's ability to absorb and stabilize the contents of essential oils, research is required to investigate novel delivery mechanisms, such as encapsulation or nano emulsion.
- *Efficacy and mechanism of action*: Although there is mounting evidence for the health advantages of essential oils, further research is needed to clarify the precise processes by which they work and to determine their effectiveness in treating a range of medical disorders. Studies conducted *in vitro* and *in vivo*, as well as clinical trials, can shed important light on the efficacy and mechanism of action of essential oils.
- *Synergistic combinations*: Examining how essential oils interact synergistically with other natural substances or prescription medications may open up new possibilities for nutraceutical compositions. Examining possible interactions and additive/synergistic effects can result in formulations that are optimized and have higher efficacy.
- *Regulations to be taken into account*: To protect consumer safety and maintain product quality, specific rules and norms regulating the use of essential oils as nutraceuticals must be established. For the labelling, dosage, and claims related to essential oil-based nutraceutical goods, regulatory organizations should set forth precise requirements.

In conclusion, while essential oils show potential as nutraceutical supplements, more study is needed to perfect their composition, comprehend their safety profiles, improve bioavailability, and establish their effectiveness in various medical diseases. The creation of safe, efficient, and standardized nutraceutical products using essential oils will be facilitated by addressing these issues.

REFERENCES

Aggarwal, B. B., Sundaram, C., Malani, N., & Ichikawa, H., 2007. Curcumin: The Indian solid gold. *Advances in Experimental Medicine and Biology*, 595, pp.1–75.

Aguirre, A., Ayala-Zavala, J.F., Silva-Espinoza, B.A. & González-Aguilar, G.A., 2020. Natural compounds from medicinal plants: Potential use as antimicrobials, antiinflammatory agents, and for the treatment of chronic diseases. *Molecules*, 25(18), p.4286.

Al-Yasiry, A.R. & Kiczorowska, B., 2016. Frankincense – therapeutic properties. *Postępy Higieny i Medycyny Doświadczalnej*, 70, pp.380–391.

Anwar, F., Ali, M., Hussain, A.I., Shahid, M. & Shah, S.A., 2019. Antioxidant and antimicrobial activities of essential oils of some aromatic plants. *Journal of the Science of Food and Agriculture*, 99(8), pp.4302–4310.

Azimi, G., Khooshnood, M., Aghaee, A. & Entezari, M.H., 2020. Nutraceutical and medicinal potentials of essential oils: A comprehensive review. *Complementary Therapies in Medicine*, 48, 102274.

Badgujar, S.B., Patel, V.V. & Bandivdekar, A.H., 2014. Foeniculum vulgare Mill: A review of its botany, phytochemistry, pharmacology, contemporary application, and toxicology. *BioMed Research International*, 2014, 842674.

Bakkali, F., Averbeck, S., Averbeck, D. & Idaomar, M., 2008. Biological effects of essential oils – a review. *Food and Chemical Toxicology*, 46(2), pp.446–475.

Barocelli, E., Calcina, F., Chiavarini, M., Impicciatore, M., Bruni, R., Bianchi, A., . . . & Ballabeni, V., 2012. Antinociceptive and gastroprotective effects of inhaled and orally administered Lavandula hybrida Reverchon "Grosso" essential oil. *Life Sciences*, 90(21–22), pp.897–903.

Baser, K.H.C., 2008. Biological and pharmacological activities of carvacrol and carvacrol bearing essential oils. *Current Pharmaceutical Design*, 14(29), pp.3106–3119.

Bhalla, Y., Gupta, V.K. & Jaitak, V., 2013. Antioxidant and cyclooxygenase (COX) inhibitory properties of essential oils from Curcuma species. *Food and Chemical Toxicology*, 62, pp.386–391.

Buckle, J., 2015. *Clinical Aromatherapy: Essential Oils in Practice* (3rd ed.). Churchill Livingstone.

Budzyńska, A., Wieckowska-Szakiel, M., Sadowska, B., Kalemba, D. & Różyło, R., 2011. Antibiofilm properties of selected plant essential oils and herbal extracts against the Escherichia coli O157:H7, Staphylococcus aureus, and Salmonella enterica Serovar Typhimurium. *Microbial Ecology*, 61(4), pp.877–887.

Burt, S., 2004. Essential oils: Their antibacterial properties and potential applications in foods – a review. *International Journal of Food Microbiology*, 94(3), pp.223–253.

Burt, S.A. & Reinders, R.D., 2003. Antibacterial activity of selected plant essential oils against Escherichia coli O157: H7. *Letters in Applied Microbiology*, 36(3), pp.162–167.

Carson, C.F., Hammer, K.A. & Riley, T.V., 2006. Melaleuca alternifolia (Tea Tree) oil: A review of antimicrobial and other medicinal properties. *Clinical Microbiology Reviews*, 19(1), pp.50–62.

Chainani-Wu, N., 2003. Safety and antiinflammatory activity of curcumin: A component of tumeric (Curcuma longa). *Journal of Alternative and Complementary Medicine*, 9(1), pp.161–168.

Choi, J.G., Kim, S.Y., Jeong, M. & Oh, M.S., 2013. Pharmacological effects of standardised extract of Siegesbeckia glabrescens on lipopolysaccharide-induced neuroinflammation. *Evidence-Based Complementary and Alternative Medicine*, 2013, pp.1–8.

Chorianopoulos, N.G., Kalpoutzakis, E., Aligiannis, N., Mitaku, S. & Nychas, G.J.E., 2004. Essential oils of Satureja, Origanum, and Thymus species: Chemical composition and antibacterial activities against foodborne pathogens. *Journal of Agricultural and Food Chemistry*, 52(26), pp.8261–8267.

Da Silva, A.C., Lopes, P.M., de Azevedo, M.M., Costa, D.C., Alviano, C.S. & Alviano, D.S., 2017. Biological activities of α-pinene and β-pinene enantiomers. *Molecules*, 22(9), p.1559.

De Cássia da Silveira e Sá, R., Andrade, L.N. & de Sousa, D.P., 2013. A review on antiinflammatory activity of monoterpenes. *Molecules*, 18(1), p.1227.

Delaquis, P., Stanich, K., Girard, B. & Mazza, G., 2002. Antimicrobial activity of individual and mixed fractions of dill, cilantro, coriander, and eucalyptus essential oils. *International Journal of Food Microbiology*, 74(1–2), pp.101–109.

Djenane, D., Yangüela, J., Amrouche, T., Boubrit, S., Boussad, N. & Roncales, P., 2012. Inhibition of Listeria monocytogenes and Staphylococcus aureus on beef by thyme essential oil and nisin in alginate edible films. *Food Control*, 23(1), pp.237–242.

Edris, A.E., 2007. Pharmaceutical and therapeutic potentials of essential oils and their individual volatile constituents: A review. *Phytotherapy Research*, 21(4), pp.308–323.

Friedman, M., Henika, P.R. & Mandrell, R.E., 2002. Bactericidal activities of plant essential oils and some of their isolated constituents against Campylobacter Jejuni, Escherichia coli, Listeria monocytogenes, and Salmonella enterica. *Journal of Food Protection*, 65(10), pp.1545–1560.

Göbel, H., Fresenius, J., Heinze, A., Dworschak, M. & Soyka, D., 2016. Effectiveness of Oleum menthae piperitae and paracetamol in therapy of headache of the tension type. *Nervenarzt*, 87(8), pp.990–998.

Göbel, H., Schmidt, G. & Soyka, D., 1994. Effect of peppermint and eucalyptus oil preparations on neurophysiological and experimental algesimetric headache parameters. *Cephalalgia*, 14(3), pp.228–234.

González-Trujano, M.E., Ponce-Monter, H.A., Martínez, A.L. & Moreno, J., 2017. Antioxidant and neuroprotective effects of Rosemary (Rosmarinus officinalis L.) extract in rodent models of Parkinson's disease. *Journal of Medicinal Food*, 20(11), pp.1039–1046.

Grigoleit, H.G. & Grigoleit, P., 2005. Peppermint oil in irritable bowel syndrome. *Phytomedicine*, 12(8), pp.601–606.

Gutierrez, J., Barry-Ryan, C. & Bourke, P., 2009. Antimicrobial activity of plant essential oils using food model media: Efficacy, synergistic potential and interactions with food components. Food *Microbiology*, 26(2), pp.142–150.

Hammer, K.A., Carson, C.F. & Riley, T.V., 1999. Antimicrobial activity of essential oils and other plant extracts. *Journal of Applied Microbiology*, 86(6), pp.985–990.

Hewlings, S.J. & Kalman, D.S., 2017. Curcumin: A review of its' effects on human health. *Foods*, 6(10), p.92.

Hill, C., Guarner, F., Reid, G., Gibson, G.R., Merenstein, D.J., Pot, B., . . . & Salminen, S., 2014. Expert consensus document: The international scientific association for probiotics and prebiotics consensus statement on the scope and appropriate use of the term probiotic. *Nature Reviews Gastroenterology & Hepatology*, 11(8), pp.506–514.

Holick, M.F., 2007. Vitamin D deficiency. *New England Journal of Medicine*, 357(3), pp.266–281.

Hussain, A.I., Anwar, F., Hussain Sherazi, S.T. & Przybylski, R., 2008. Chemical composition, antioxidant and antimicrobial activities of basil (Ocimum basilicum) essential oils depends on seasonal variations. *Food Chemistry*, 108(3), pp.986–995.

Juergens, U.R., Dethlefsen, U., Steinkamp, G., Gillissen, A., Repges, R. & Vetter, H., 2003. Antiinflammatory activity of 1.8-cineol (eucalyptol) in bronchial asthma: A double-blind placebo-controlled trial. *Respiratory Medicine*, 97(3), pp.250–256.

Juergens, U.R., Stöber, M. & Schmidt-Schilling, L., 1998. Antiinflammatory effects of eucalyptol (1.8-cineole) in bronchial asthma: Inhibition of arachidonic acid metabolism in human blood monocytes ex vivo. *European Journal of Medical Research*, 3(9), pp.407–412.

Kamatou, G.P., Viljoen, A.M. & Steenkamp, P., 2013. Antioxidant, antiinflammatory activities and HPLC analysis of South African Salvia species. *Food Chemistry*, 136(2), pp.466–472.

Khanna, R., MacDonald, J.K. & Levesque, B.G., 2012. Peppermint oil for the treatment of irritable bowel syndrome: A systematic review and meta-analysis. *Journal of Clinical Gastroenterology*, 46(8), pp.1–6.

Kim, D.S., Kim, J.S., Jeon, Y.D., Han, Y.H., Keum, K.S. & Chung, H.T., 2020. Citral exerts antiinflammatory and anti-oxidative effects through the MAPK/Nrf2 pathway in lipopolysaccharide-stimulated RAW264.7 cells. *Biomolecules*, 10(3), p.462.

Koulivand, P.H., Khaleghi Ghadiri, M. & Gorji, A., 2013. Lavender and the nervous system. *Evidence-Based Complementary and Alternative Medicine*, 2013, pp.1–10.

Kuroda, K., Inoue, N., Ito, Y., Kubota, K., Sugimoto, A., Kakuda, T. & Fushiki, T., 2005. Sedative effects of the jasmine tea odor and (R)-(−)-linalool, one of its major odor components, on autonomic nerve activity and mood states. *European Journal of Applied Physiology*, 95(2–3), pp.107–114.

Lawless, J., 2013. *The Encyclopedia of Essential Oils: The Complete Guide to the Use of Aromatic Oils in Aromatherapy, Herbalism, Health, and Well Being*. Conari Press.

Liao, L.Y., He, Y.F., Li, L. & Meng, H., 2019. Scutellaria baicalensis Georgi: A review of traditional uses, phytochemistry, pharmacology and toxicology. *Journal of Pharmacy and Pharmacology*, 71(9), pp.1353–1369.

Lis-Balchin, M., 2006. *Aromatherapy Science: A Guide for Healthcare Professionals. Pharmaceutical Press.*

Moss, M., Cook, J., Wesnes, K. & Duckett, P., 2003. Aromas of rosemary and lavender essential oils differentially affect cognition and mood in healthy adults. *International Journal of Neuroscience*, 113(1), pp.15–38.

Moss, M., Hewitt, S., Moss, L. & Wesnes, K., 2008. Modulation of cognitive performance and mood by aromas of peppermint and ylang-ylang. *International Journal of Neuroscience*, 118(1), pp.59–77.

Mozaffarian, D. & Rimm, E.B. 2006. Fish intake, contaminants, and human health: Evaluating the risks and the benefits. *JAMA,* 296(15), pp.1885–1899.

Nazzaro, F., Fratianni, F. & Coppola, R., 2013. Essential oils and antifungal activity. *Pharmaceuticals,* 6(12), pp.1543–1575.

Peana, A.T., D'Aquila, P.S., Panin, F., Serra, G., Pippia, P. & Moretti, M.D., 2002. Anti-inflammatory activity of linalool and linalyl acetate constituents of essential oils. *Phytomedicine,* 9(8), pp.721–726.

Perry, N.S., Houghton, P.J., Theobald, A., Jenner, P. & Perry, E.K., 2000. In-vitro inhibition of human erythrocyte acetylcholinesterase by Salvia lavandulaefolia essential oil and constituent terpenes. *Journal of Pharmacy and Pharmacology*, 52(7), pp.895–902.

Perry, N.S. & Perry, E.K., 2006 Aromatherapy in the management of psychiatric disorders: Clinical and neuropharmacological perspectives. *CNS Drugs*, 20(4), pp.257–280.

Prakash, B., Kedia, A., Mishra, P.K. & Dubey, N.K., 2015. Plant essential oils as food preservatives to control moulds, mycotoxin contamination and oxidative deterioration of agri-food commodities – Potentials and challenges. *Food Control*, 47, pp.381–391.

Price, S. & Price, L., 2012. *Aromatherapy for Health Professionals* (4th ed.). Churchill Livingstone.

Prusinowska, R., Śmigielski, K. & Śmigielska, A., 2019. Composition, biological properties and therapeutic effects of lavender (Lavandula angustifolia L.): A review. *Herba Polonica*, 65(3), pp.61–78.

Rao, P.V. & Gan, S.H., 2014. Cinnamon: A multifaceted medicinal plant. *Evidence-Based Complementary and Alternative Medicine,* 2014, 642942.

Ribeiro-Santos, R., Andrade, M., Melo, N.R. & Barbosa, M., 2018. Essential oils in food preservation: Mode of action, synergies, and interactions with food matrix components. *Frontiers in Microbiology,* 9, p.1–14.

Roopesh, M.S. & Padmakumar, K., 2015. Microencapsulation of essential oils for improving their application in food and agricultural products. *Journal of Food Science and Technology*, 52(7), pp.4329–4341.

Sadiki, M., Taktak, M. and Kouisni, L., 2017. Chemical composition and antioxidant activity of essential oil of lemon (Citrus limonum) fruit peels. *Asian Journal of Chemistry*, 29(2), pp.353–356.

Sasannejad, P., Saeedi, M., Shoeibi, A., Gorji, A., Abbasi, M. & Foroughipour, M., 2012. Lavender essential oil in the treatment of migraine headache: A placebo-controlled clinical trial. *European Neurology,* 67(5), pp.288–291.

Satou, T., Kasuya, H., Maeda-Yamamoto, M., Onda, K., Tokunaga, O. and Honda, G., 2019. Antiinflammatory effects of rosmarinic acid in experimental allergic rhinitis. *Molecular Medicine Reports,* 19(6), pp.4645–4654.

Sayorwan, W., Ruangrungsi, N., Piriyapunyporn, T., Hongratanaworakit, T. and Kotchabhakdi, N., 2013. Effects of inhaled rosemary oil on subjective feelings and activities of the nervous system. *Scientia Pharmaceutica,* 81(2), pp.531–542.

Serafini, M., Peluso, I. & Raguzzini, A., 2010. Flavonoids as antiinflammatory agents. *Proceedings of the Nutrition Society,* 69(3), pp.273–278.

Serafino, A., Sinibaldi Vallebona, P., Andreola, F. and Zonfrillo, M., 2008. Stimulatory effect of Eucalyptus essential oil on innate cell-mediated immune response. *BMC Immunology*, 9, p.17.

Setzer, W.N., 2009. Essential oils and anxiolytic aromatherapy. *Natural Product Communications*, 4(9), pp.1305–1316.

Seyedian, S.S., Naseri, M., Bakhtiarian, A., Movahed, A., Tabatabaee, S.M., Yazdanparast, E. and Mohammadi, M., 2020. Comparison of the effects of lavender and tramadol on postoperative pain, current perception threshold, and mood: A randomized, double-blind clinical trial. *Journal of Research in Medical Sciences*, 25, pp.54.

Shahidi, F. & Ambigaipalan, P., 2015. Phenolics and polyphenolics in foods, beverages and spices: Antioxidant activity and health effects – A review. *Journal of Functional Foods*, 18, pp.820–897.

Shan, B., Cai, Y.Z., Sun, M. and Corke, H., 2018. Antioxidant capacity of 26 spice extracts and characterization of their phenolic constituents. *Journal of Agricultural and Food Chemistry*, 56(9), pp.3512–3519.

Shrivastava, R., Pechadre, J.C. & John, G.W., 2010. Tanacetum parthenium and Salix alba (Mig-RL) combination in migraine prophylaxis: A prospective, open-label study. *Clinical Drug Investigation*, 30(3), pp.143–153.

Siddiqui, M.Z., 2011. Boswellia serrata, a potential antiinflammatory agent: An overview. *Indian Journal of Pharmaceutical Sciences,* 73(3), pp.255–261.

Silva, J., Abebe, W., Sousa, S.M., Duarte, V.G., Machado, M.I. and Matos, F.J., 2003. Analgesic and antiinflammatory effects of essential oils of Eucalyptus. *Journal of Ethnopharmacology*, 89(2–3), pp.277–283.

Singh, A. and Kumar, S., 2019. Applications of essential oils in food preservation: A review. *Journal of Food Science and Technology*, 56(2), pp.486–497.

Skandamis, P.N. and Nychas, G.J.E., 2001. Effect of essential oil from oregano (Origanum vulgare L.) on survival and growth of Salmonella enteritidis and Escherichia coli O157:H7 on fresh-cut lettuce. *Journal of Food Protection*, 64(7), pp.1012–1018.

Soković, M., Glamočlija, J., Marin, P.D., Brkić, D., van Griensven, L.J. and van Klink, J.W., 2009. Antibacterial effects of the essential oils of commonly consumed medicinal herbs using an in vitro model. *Molecules*, 14(1), pp.4160–4171.

Soković, M., van Griensven, L.J.L.D., Brković, D. and Dajić-Stevanović, Z., 2010. Antimicrobial activities of essential oils and their components against the three major pathogens of the cultivated button mushroom, Agaricus bisporus. *European Journal of Plant Pathology*, 126(1), pp.57–66.

Teixeira, B., Marques, A., Ramos, C., Neng, N.R., Nogueira, J.M.F., Saraiva, J.A. and Nunes, M.L., 2013. Chemical composition and antibacterial and antioxidant properties of commercial essential oils. *Industrial Crops and Products,* 43, pp.587–595.

Tisserand, R. and Young, R., 2014. Essential Oil Safety: A Guide for Health Care Professionals, 2nd ed. Churchill Livingstone.

Viljoen, J., Camm, E.J. and Veale, D.J., 2012. A systematic review and evaluation of the efficacy of aromatherapy in treating symptoms related to attention deficit/hyperactivity disorder. *Journal of Alternative and Complementary Medicine,* 18(6), pp.527–536.

Wang, S.M. and Kulkarni, L., 2020. A Review of Aromatherapy as a Potential Treatment for Psychiatric Disorders. *The Mental Health Clinician*, 10(4), pp.245–256.

Weaver, C.M. and Plawecki, K.L., 1994. Dietary calcium: Adequacy of a vegetarian diet. *The American Journal of Clinical Nutrition*, 59(5), pp.1238S–1241S.

Index

Note: Page numbers in *italics* indicate a figure on the corresponding page.

For Product Safety Concerns and Information please contact our EU representative GPSR@taylorandfrancis.com Taylor & Francis Verlag GmbH, Kaufingerstraße 24, 80331 München, Germany